Along the
RIVERBANK
THE LIVING COUNTRYSIDE

A Reader's Digest selection

ALONG THE RIVERBANK

First Edition Copyright © 1985
The Reader's Digest Association Limited, Berkeley Square House,
Berkeley Square, London W1X 6AB

Reprinted with amendments 1993

Copyright © 1985
Reader's Digest Association Far East Limited
Philippines Copyright 1985
Reader's Digest Association Far East Ltd

Originally published as a partwork,
The Living Countryside
Copyright © 1981, 1982, 1983, 1984
Eaglemoss Publications Ltd and Orbis Publishing Ltd

® READER'S DIGEST, THE DIGEST and the Pegasus logo
are registered trademarks of
The Reader's Digest Association Inc. of Pleasantville, New York, USA.

ISBN 0-276-39294-9

Front cover picture: Summertime on the River Severn – lush vegetation
lines the banks right down to the water's edge.

Along the RIVERBANK

THE LIVING COUNTRYSIDE

PUBLISHED BY THE READER'S DIGEST ASSOCIATION LIMITED
LONDON   NEW YORK   MONTREAL   SYDNEY   CAPE TOWN

Originally published in partwork form
by Eaglemoss Publications Limited and Orbis Publishing Limited

Consultant

Nigel Holmes

Contributors

Contents

Along the
RIVERBANK

Introduction

Most of us have a popular vision of a river. The image we conjure up of its source usually includes windswept mountains and a narrow cascading stream tumbling past boulder-strewn moorland. The picture changes as the stream matures into a river; dropping through grazed uplands, its mood can change from a relaxed descent alongside guardian trees and fringing herbs to one of youthful exuberance over tumbling rock falls. On the lowland approaches to the estuary, the idealised river takes on a soporific air amid weeping willows and waterlilies as senility and salinity approach in a seemingly aimless meander to the sea.

Alas, few rivers conform to our dreams. No rivers in south-east England have sources with highly effervescent characteristics, while few north of the Border have relaxed reedy lowland reaches. It is for this reason that the plant and animal communities of rivers in different parts of the country vary so much. The flashing, cool mountain stream is the home of lowly algae, mosses and liverworts, streamlined invertebrates requiring highly oxygenated water, spawning salmon and bobbing dippers. In contrast, the lowland river favours colourful flowers, coarse fish, moorhens nesting on reed-fringed banks, foraging water voles and damselflies skimming over warm water surfaces. Between them may be waterfalls, earth-cliffs, shingle islands and many other environments.

Each supports its own blend of wildlife – the greater the variety of habitats in a river, the greater is the diversity of plants and animals. Rivers, and the connecting mosaic of habitats which constitute the corridor through which they flow, are thus the most important 'linear' habitat in Britain.

Left: Balnagarg Burn in Scotland – a clean, highly oxygenated, fast-flowing mountain stream running over rock.

RIVERS AND THEIR HABITATS

Every river has its own distinct character and follows a course through a series of habitats that merge almost imperceptibly as the water flows from source to sea.

The beginnings of a river are often inauspicious. Even the mighty Thames starts in a damp patch in a field near Cirencester, and the Wye and Severn rise close together in a bog below the mountain of Plynlymmon in Wales. Most rivers rise on high ground and change in character as the gradient lessens, the current slackening and the bed widening and deepening. Finally they pass through a brackish estuary into the sea.

The conditions of life in the various stages of running water differ considerably from those in the still waters of ponds and lakes. The most important factor which determines the kinds of plants and animals that can live in rivers is the speed of the current. This affects not only the turbulence and ever-present danger of being swept away, but determines the nature of the bed over which the river flows. It also has some effect on the amount of oxygen in the water for respiration.

Another factor is the amount of dissolved salts which the streams and gullies that flow into a river have drained out of the surrounding rock and soils. Water plants need the familiar fertilizers that we use in our gardens: potassium, nitrates and phosphates. Calcium, in the form of calcium carbonate or lime, is needed by molluscs for strengthening their shells. Crustaceans such as freshwater shrimps and crayfish also need it for strengthening the hard external covering of their bodies.

In a river, unlike a pond or lake, there can be little accumulation of nutrients, or recycling of dead organic material, because everything is carried away by the constant flow of water.

Upper course A number of small, almost imperceptible trickles of water may have to join up before it is clear that a stream is in the making. The water is shallow and the vegetation through which it runs is chiefly mosses, especially sphagnum. Even in this unpromising habitat there are myriads of minute animals among the mosses: rotifers, water-bears and protozoans related to *Amoeba* but with shells. Most of these creatures feed on small particles of plant remains or algae on

Above: The Duddon river in Cumbria clearly shows the three main stages in a river and the kind of habitats through which it passes from its source to the sea. Here the upper course is little more than a mountain stream, in which little grows and only a restricted number of animal species can survive.

Right: The freshwater shrimp, which survives in the upper course of rivers, is an active swimmer and therefore can avoid being swept downstream with the current. Its body is unusually flattened and crawling into small crevices between stones presents no difficulties.

the mosses. Water-bears pierce the cells of the mosses with their mouthparts and suck out the contents.

Turn over a stone and you could find black planarian worms crawling over the surface and perhaps the stony cases of caddis fly larvae. In the damp earth at the edge of the water you may find the large 'leather-jacket' larvae of crane-flies such as *Pedicia* which feed on smaller fly larvae. One of the snails of these damp places is the dwarf pond snail, the host of the early stage of the liver-fluke which causes serious losses among sheep and cattle grazing on upland pastures.

Eventually, a number of small head-streams meet to form a wider, deeper channel with a more constant and rapid flow. If the gradient is steep, there may be small waterfalls or splashes which assist aeration of the water. All but the largest stones are swept along by the current, leaving the bottom rocky with perhaps coarse gravel in less turbulent places such as the bend of a stream. Rooted plants are unable to gain a foothold, but the stones are covered with algae and here and there clumps of mosses and liverworts.

Sweeping currents These conditions are ideal for animals that need cool and well-oxygenated water, provided they can cope with the strong current and avoid being swept downstream. It is no place for the insects that breathe atmospheric air and have to rise to the surface to obtain it; they would soon be swept away. Here gills are necessary to extract some of the rich supply of dissolved oxygen in the water; these are possessed by the nymphs of mayflies, stoneflies and caddis flies that live in this part of the river. Most live under the stones, during the daytime at least, and browse on the algae. Some, such as the larvae of the mayfly called by anglers the great red spinner, have very flat bodies to minimise surface resistance to the current. All have long claws on their legs to hold firmly to the rough surfaces of stones.

If dislodged, stream animals swim upstream and regain their foothold as soon as they can. In this way they do not get swept very far from the habitat that suits them best. If they went with the current they might be carried too far from their favoured part of the river.

Using stones Instead of the plant material which more familiar caddis fly larvae use to make their protective cases in ponds, stream caddis have to use small stones. Some attach the cases to the underside of larger stones while browsing and when they pupate.

If you turn over the stones at the edge of streams you will see hundreds of the little oval, flat cases of *Agapetus* and perhaps also the long, cylindrical ones of *Potamophylax*, firmly fixed to the underside. *Silo* and *Goera* use pebbles as ballast to minimise the risk of being swept away. Others make do with shelters of silk among the stones while *Hydropsyche*, the 'spider of the rapids', spins a funnel-shaped net in which it traps food particles swept down by the current.

Some stones in the fastest current become covered with hundreds of tiny larvae of the *Simulium* fly. They spin a silken mat on the stone to which they attach their rear end, while the front end waves freely in the current. Two huge fans of bristles on the head sift out food particles from the water.

Suckers are efficiently used by some stream animals to avoid being swept away. The little river limpet can move about freely on exposed stones using its broad muscular foot as a sucker. Additionally, the edge of the shell is soft and fits neatly into irregularities on stones. The smooth, mucus-covered under-surface of planarian worms enables them to fit closely to rock surfaces when crawling about. Leeches have suckers at both ends of their bodies, the larger one at the rear end. Using them alternately, they loop safely over stones.

Fish population The characteristic fish and ultimate predator of this section of river is the brown trout, a species which needs well-oxygenated water and a cool temperature to thrive. Another fish found in this section is the small miller's thumb or bullhead, so called from the large head which is flattened and enables the fish to creep under stones. The stone loach may also be found here. This longer and more cylindrical fish is easily distinguished by the two large and four small barbels under the mouth. Both species feed on small invertebrates on the bottom.

Middle course As the river reaches flatter country the current slackens and the volume of water increases. Temperatures fluctuate more than in the upper reaches, but there is still enough oxygen for some of the animals already mentioned – even the trout, although it migrates upstream in winter to spawn.

It is a surprising fact, only recently realised, that drifting matter including dead leaves and other plant material that has fallen or been blown into the water, is the major food of river animals – and not the large aquatic plants themselves (although the algae growing on them are important in the river's economy). Leaves and bits of plants are soon broken down by bacteria and microscopic

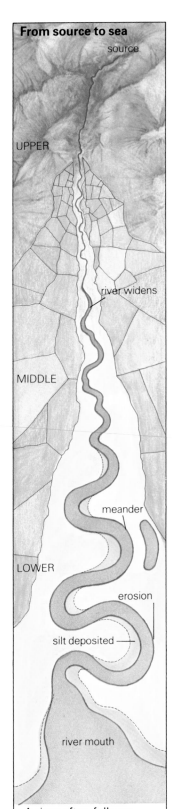

From source to sea

source

UPPER

river widens

MIDDLE

meander

LOWER

erosion

silt deposited

river mouth

A river often follows three main stages. In the upper course it starts as a fast mountain stream, cascading over rocks. As it flows into lower-lying land it widens, eroding the banks in places. In its lower reaches the slow moving water deposits silt and occasionally floods the flat land as it meanders towards the sea.

fungi to provide readily digestible food for insect nymphs, snails and the real scavengers, freshwater shrimps and water slaters.

The middle course of the river is sometimes called the minnow-reach, for this beautiful little fish feeds in silvery shoals in the shallows. It takes any small animals it can find, although it is not averse to plants.

Lower course Widening and deepening, the river continues its course over flatter country, now with a much reduced flow. It deposits silt, especially at bends and in bays cut out of banks during winter floods. Rooted plants of many kinds can become established, including water-lilies, Canadian waterweed and water starwort. Conditions at this stage are comparable to those in ponds. Variable temperatures and fluctuating oxygen levels mean that animals of the upper sections cannot survive here.

One mayfly, however, is common in the sandy mud at the margins; this is the burrowing nymph of the largest mayfly *Ephemera*, the mayfly of the fly fisherman. In early June huge groups of the emerging males gather into mating swarms, rising and falling in unison. Females which join them are fertilized and then deposit their eggs in the water. Shortly after, both sexes become the 'spent gnat' of the angler as they float dying on the surface.

In chalk and limestone streams lives Britain's largest freshwater crustacean, the freshwater crayfish. It is similar in appearance to the marine lobster, but only about 10-15cm (4-6in) long. During the day it lives in burrows in the bank and at night emerges to search for food. In spite of the crayfish's huge pincer claws, which can catch and hold quite large animals, it often eats plant material.

Pond-like conditions In the sheltered areas among the plants many familiar animals more typical of ponds find refuge. On the surface the pond skaters, water crickets and whirligig beetles are abundant. Below the surface, water boatmen and other water bugs, water beetles of several species, dragonfly and damselfly nymphs, leeches and water snails

Above: A section of the middle course of the River Duddon. With a slower current and more water, silt is deposited and a few plants, such as the water-crowfoot (below) and the water starwort, can find footholds. Both can resist the pull of the current, their long stems streaming out in the flow. Willow moss, our largest moss, is also at home here, firmly anchored to stones.

can be seen. In lime-rich rivers you may find freshwater winkles. Unlike pond snails which breathe atmospheric air with lungs, these have gills and can extract dissolved oxygen from the water.

The larvae of the midges of the huge family Chironomidae occur at the river edge, in vast numbers in or on the mud in silken tunnels covered with debris. The reddish colouring of some of them indicates that their blood contains haemoglobin to enable them to absorb oxygen from surroundings where it is relatively scarce. This also applies to another mud inhabitant, the sludge worm.

On stones or wooden posts at the edge of clean rivers, encrusting masses of river sponge occur, sometimes green in colour because of the presence of minute algae. In this symbiotic relationship (one in which each participant gains something from the other), the algae get nutrients from the excretions of the sponges, which in turn obtain oxygen from the algae's photosynthesis.

Partly buried in the mud at the margins of the river live the large freshwater mussels, either the swan or duck mussel. Their hinged shell is slightly opened, with two tubes. The intake siphon takes in a current of water and, after extracting the oxygen and food particles, the exhalant siphon expels it.

Characteristic fish are the members of the carp family – carp, chub, tench, roach and bream. But water authorities and angling clubs may stock the rivers with other species as well.

Approaching the sea As the river finally approaches the sea and comes within reach of tides violent fluctuations in salinity, temperature and current make life difficult for animals used to freshwater conditions. One or two can cope, including a few beetles such as *Platambus* and *Hydrobius*.

Do not be misled by what seem to be familiar animals. Pond skaters may be on the surface, but they will be different species from those further up the river, such as *Gerris thoracicus* rather than *G. najas*. The freshwater shrimps will be *Gammarus zaddachi* instead of *G. pulex*.

Among fish, the river lamprey spends most of its time in the sea and only comes into rivers in autumn to spawn. So do salmon and sea trout (the migratory variety of the common brown trout).

The future Few rivers are now unaffected by man. Enrichment of some stretches by the nitrates and phosphates of agricultural chemicals washed into the water can cause excessive growth of plant life, especially algae. When this dies down the oxygen used in decomposition may seriously reduce the amount available to animals and some die.

Lower reaches of rivers near centres of population may be similarly affected by the effluents from sewage treatment plants, even those which use modern treatments.

The mud surface at the edge of rivers in the lower course is often pitted with small holes out of which emerge hundreds of waving reddish tails. These belong to sludge worms (seen above as a balled mass for use as angler's fish bait). Their front end is buried in the mud ingesting particles of food while the tails extract oxygen from the surrounding water. They survive in these conditions because the flow of water is much reduced, as on this stretch of the Duddon (below).

STREAMS THAT RISE FROM CHALK

Although they are often thought of as natural and untouched by man, the chalk streams that occur in southern and eastern England are the product of deforestation, farming, water supplies, weed cutting and artificial stocking with fish. They are, in consequence, very different from their original state.

Below: The River Itchen (Hampshire) in June, with masses of water-crowfoot growing up to the surface. On some chalk streams these and other weeds are cut twice yearly. The fertile soils of chalk streams have always provided ample nutrients for plants – at present over 40 times as much nitrogen, phosphorus and potassium flow past the water plants as they need for their annual growth.

The soft, calcareous, porous rock known as chalk only occurs in England, France and New Zealand. In England it is found in hills (downs) and plains within a crescent extending from Dorset, through Kent and Hertfordshire to Norfolk and North Humberside. The smaller rivers rising from this rock are known as chalk streams. As a result of a rather special combination of climate, geology and human activity, they have several characteristic features which encourage plant and animal life and make them unique.

Natural characteristics Most of the water of chalk streams comes from rainfall which has percolated slowly through the overlying soil and accumulates in the chalk. Hence it is stable in temperature, clear, and rich in calcium, nitrates and carbon dioxide. Also, downstream of the permanent springs, variations in flow are small.

Further up the valleys are water-courses, known as winterbournes, which flow only in the winter and spring when the water levels in the chalk are at their highest. The valleys are neither particularly steep nor particularly flat, so the water flow is moderately fast. This fast flow helps to keep the water well aerated, prevents the accumulation of large areas of silt, brings a continual supply of mineral nutrients and organic food to the river plants and animals and rarely causes damage.

Man's activities The features described above have always been the natural characteristics of chalk streams, but many others have arisen from man's activities. The early history of the lower valleys of the River Frome in Dorset, for instance, is known from pollen and other remains preserved in peat which has accumulated in permanent reed swamps since the end of the Ice Age.

For nearly 10,000 years after the Ice Age the valleys were thickly wooded, with pine and later oak on the sides and alder or willow in the flood plains. The rivers are likely to have flowed in several small channels in deep shade through these forests. Water plants would have been few and most animal food would have come from tree leaves. Many of the animals common now, which depend on water plants for food, support and shelter,

Above: Dense growths of water weeds are typical of chalk streams. In the smaller streams water-cress (shown here), fool's water-cress and lesser water-parsnip are common, while in the larger streams water-crowfoot is usually the dominant species.

Below: An old water mill at Longparish on the River Test in Hampshire. Such mills were first built by the Romans, and then by the Saxons. They altered the speed of water flow and consequently the plants that were able to grow in the water. Although very few of the mills work today, the altered habitats remain.

would have been scarce then. For example, the grannom fly and many of the reed smuts or blackflies would not have been abundant.

Forest clearance was started around 1000 BC and many of the valley bottoms were cleared by Roman times, which encouraged the growth of water plants. The Romans and Saxons started building water mills, small at first but later much larger. In the basin of the River Frome there were over 50 such mills. Each had a weir, a mill-leat and a tail-race (a watercourse to and from a mill), which introduced new types of habitat with either unusually fast or unusually slow water flows. Although very few of these now work, the altered habitats remain. In the slow waters pondweeds began to grow. At first the perfoliate pondweed appeared and later, after its introduction from North America, Canadian pondweed became important. Another species, *Elodea nuttallii*, has recently been introduced. Typical animals, not common in normal streams, now include the water skater, pond olive, damselflies and roach.

A little later drainage of the fields in the valley bottoms started, reaching a climax between 1650 and 1900 in the elaborate system of water meadows. These took much of the flow into new or deepened and narrowed channels. The damp, grazed meadows became rich in flowers, including several species of marsh orchid, and marsh cinquefoil.

As the populations of the valleys increased, human and animal sewage affected the chalk rivers (nowadays it is mostly discharged as well-treated mineralised effluents), increasing the amount of phosphorus entering the water. Recently agricultural fertilisers have also increased the nitrogen entering the rivers. However, the naturally fertile soils of the chalk streams have probably always provided ample nutrients for plant life.

Many of the springs are used as water-cress beds or trout farms, these uses modifying the temperature, mineral and organic content of the water, and on occasions plants or non-native fish (such as the rainbow trout) are released into them.

More and more subterranean water is abstracted from the chalk for man's use,

Above: A brown trout swimming among water-starwort above the gravelly bottom of the River Test. The chalk stream gravel beds, abundance of water weeds and waters flowing at different speeds are well suited to the lifestyle of this fish. Trout, and even salmon, are sometimes artificially introduced into the streams, making the fishing rights extremely valuable.

Left: The kingfisher is drawn to chalk streams by such fishes as the bullhead.

Above: A mayfly just emerged from its aquatic nymphal stage. The usual brief lifespan of adult mayflies — less than a day — is often even shorter.

which has sometimes decreased the flow. This may lead to a reduction in water-crowfoot and animals such as the mayfly and trout, while increasing silt, silt-loving animals and blanket-weed. However, recent developments put water pumped from the chalk into the streams, so that it flows downstream to the towns requiring it. This is often done in times of low flow, thus improving the river.

Chalk stream plants Dense growths of water weeds are typical of chalk streams. In the smaller streams you can find fool's water-cress, lesser water-parsnip or water-cress while in the larger streams chalk-stream water-crowfoot is usually dominant. These weeds hinder the flow of water and cause spring and summer flooding. To prevent flooding, and to make angling easier, these weeds have been cut once or twice a year for at least 150 years at many sites, probably for much longer. In the past this was done selectively by gangs of men with scythes, but now most Water Authorities use mechanical cutters in the larger channels.

The intensity, manner and frequency of weed cutting greatly influence water life. Water-crowfoot, in particular, appears to thrive on the normal pattern and it would change its abundance if cutting was altered. If cutting is stopped, the amount of weeds present in the summer decreases in the following years. An unusually heavy and late cut has been shown to decrease water-starwort

and increase water-crowfoot in the next year.

Some fish, such as roach, spawn on bankside weeds near the surface about the same time as the first weed cut. A fall in water level following an early cut could kill many of their eggs. Mortalities of young fish are higher in cut reaches of rivers, and in small streams herons can soon catch all the trout when there is little weed left. Water-cress has very dense populations of water-shrimps and non-biting midges, which are reduced or lost when the plant is absent.

In the winterbournes there are plant species which can tolerate dry periods because they have aerial leaves and deep roots, and which thrive in the absence of management. Fool's water-cress is one example.

Thriving fish Chalk streams are particularly suitable for trout. There are gravel beds for spawning and weed beds and a variety of currents to provide cover, territories and a plentiful supply of food. The stable water temperatures provide trout with the cool water they need in summer and the warmer spring and autumn temperatures allow early hatching and a high growth rate. The fishing rights are very valuable, especially when salmon are present, and many of the fish populations are managed. When anglers control the management, weed cuts are made to encourage fish food such as water-shrimps and flies and their nymphs. Gravel beds for spawning are encouraged in many streams.

Coarse fish, especially grayling and the predatory pike are discouraged or removed (50,000 a year in some places). Nevertheless, other fish often remain biologically more important and productive. In the smaller streams these are the bullhead and the minnow and, in larger streams, eels and dace.

Chalk streams
1 River Hull (upper reaches above Beverley)
2 River Bain
3 River Nar
4 River Lea (above Chingford)
5 River Chess
6 River Lambourn
7 River Kennet
8 River Itchen
9 River Test
10 River Avon (Wiltshire)
11 River Wylye
12 River Frome
13 River Allen
14 River Great Stour
The lower reaches of these rivers may not have all chalk stream features.
The rivers Witham, Great Ouse and Thames connect the chalk streams that flow into them but are not chalk streams themselves. (These are marked on the map with a dotted line.)

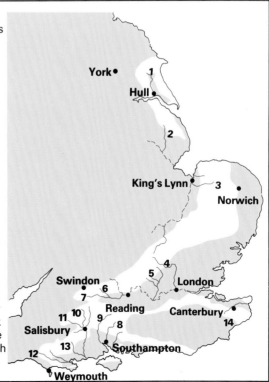

CANALS: HAVENS FOR WILDLIFE

Britain's 3000 miles of canals – man-made waterways built mostly in the 18th and 19th centuries to carry commercial freight – provide a stable wetland habitat for wildlife.

Although the Romans dug Britain's first canal – the Fossdyke – in the second century AD, the great period of artificial waterway construction occurred between the middle of the 18th and the middle of the 19th centuries. These canals, constructed for the carriage of freight between the mining, industrial and commercial centres of the country, linked much of the navigable river system of Britain with a network totalling over 3000 miles in length.

The commercial heyday of the canals was short-lived. The railways took over from the horse-drawn narrow boats and, by the start of the 20th century, canal traffic was declining rapidly and losing its financial

viability. As the decline continued many of the smaller branches of the canal network fell into total disuse, but after World War II the greater part of the canal system was included in the programme of nationalisation and came under the control of the Ministry of Transport. In 1962 the British Waterways Board was set up by Act of Parliament to take responsibility for the canals.

Today commercial traffic is almost non-existent, but its place has been taken by a massive upsurge in recreational use – not only allowing the larger canals to be kept open, but also enabling sections of disused ones to be brought back to a working state. If the mileage of canals still in working use is added to that of those which have been abandoned, there is what is virtually a wildlife sanctuary some 3000 miles long.

Canal structure Although a canal may seem to have much in common with a river as far as its wildlife is concerned, there are important structural and physical differences that have considerable bearing on its ecology. In contrast to a river, with its deep, calm pools and rapid-flowing shallows, the variable nature of its bed and a marked fluctuation in water level, a canal is more or less of uniform profile throughout its length. It is deepest in the middle and shelves gradually to shallower margins; there is very little fluctuation in water level, and what is perhaps of greatest consequence to the ecology, there

Above: A stretch of the Grand Union Canal, North Kilworth. This is a tidy, 'managed' canal but its banks and towpath support a wide range of plants – with their associated insect, bird and animal life.

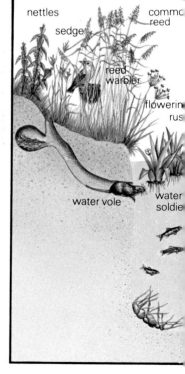

nettles
commo reed
sedge
reed warbler
flowerin rus
water vole
water soldie

is little flowing movement of the water. A canal is spared the scouring, eroding effect of a river's spate.

The amount of water allowed into a canal is only sufficient to maintain the water level. A controlled flow of water–coming mainly from adjacent rivers or streams– enters the canal at its highest point and flows to its lowest point through a series of locks. This means that noticeable flow occurs only when a lock is opened. The mineral content of the water is influenced to a limited degree by the soil of the retaining banks, while the flowers and plants that grow along the banks are a guide to the alkalinity or otherwise of the soil.

The large surface area of a canal in relation to its depth means that during the long days of summer the water temperature rises considerably, encouraging prolific phyto-plankton and algal growth–sometimes to such a degree that the water appears green. The abundance of animal life, from the microscopic zooplankton to the larger insects, crustaceans and molluscs ensures a habitat in which fish can thrive.

Canal ecology There are three ecological zones in a canal: the navigation channel, which is the deepest part, the shallower marginal areas, and the retaining banks. The line of separation between these is not always easy to define: for instance, the marginal shallows zone may be absent, where the banks are reinforced with steel or concrete piling, for example.

Only a few plants are found in the navigation channel and these are not permitted to develop to an extent where they would hinder navigation. A number of fish occur here, from the rarely-seen, bottom-feeding tench to sizeable bream, carp and roach. The

Above: Part of the old Basingstoke Canal, when it was choked with water soldier (*Stratiotes aloides*). As the water level in a disused canal falls, the marginal vegetation extends outwards from the banks and the central channel becomes narrower and shallower. With the changing environment new species become established and in a few decades the waterway is choked with vegetation. The habitat is at its richest at this stage.

Right: Indian balsam (*Impatiens glandulifera*). This, and another species, the orange balsam (*I. capensis*), are both introduced plants that have spread widely along our canal network.

The three canal zones

Ecologically a canal is made up of three zones: the navigation channel, which has the greatest depth of water; the marginal shallows; and the banks and towpath. The navigation channel contains the least number of plant and animal species and is the part most affected by the passage of boats. The marginal shallows are characterised by a large number of plant species and insects. The banks of a canal are inhabited by many mammals and birds and have a luxuriant plant growth.

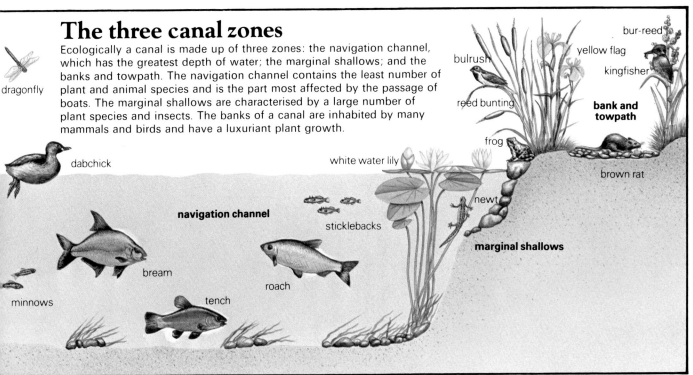

Lancaster canal, in particular, has an astonishing variety of fish species.

A large number of plant species grows in the marginal shallows zone. Floating plants include the duckweeds (*Lemna* species), while pondweeds such as *Potamogeton crispus* have totally submerged foliage.

Undoubtedly the most abundant flora of the shallows are the emergent plants, with foliage above the surface. These plants provide cover for nesting birds that find seclusion and relative safety among the luxuriant growth of bur-reed, reedmace and common reed.

In this area of shallow water several less common plants occur. Perhaps the loveliest of them all is the flowering rush, a plant characteristic of our canal margins. Among several introduced plants well-established as members of the waterway environment is the curious-looking sweet flag (*Acorus calamus*), a plant not often noticed when its strange spike of greenish flowers is not in bloom. A more recent introduction is the orange balsam or jewel-weed, but this has spread so widely along our canal network that it can be regarded as the plant most typical of the canal habitat.

Many of the insects of the waterways can be found among the vegetation of the shallows. Dragonfly nymphs are not easy to see, but the adult dragonflies attract attention by their size, dashing flight and gaudy colours. Canals are particularly good habitats for aquatic snails; the steady water flow brings a regular supply of nutrients, there are no sudden catastrophes such as floods which could sweep the snails away, and the temperature is generally warm.

High banks are not often seen in canals, but where they occur they are often riddled with holes made by the water vole, perhaps

Above: Where a canal towpath is no longer in constant use, and under wear from the feet of walkers and anglers, plants such as the stately yellow flag (*Iris pseudacorus*) can become established.

Opposite page: The easily identifiable coot is one of many birds to frequent canals in spring and summer.

Below: Peak Forest Canal at Lyme Green, Macclesfield. A period of prolonged and severe frost has covered the canal with a sheet of ice. When a canal ices over like this, birds such as coots and moorhens which are dependent on open water, must move elsewhere, but the ice has little effect otherwise on life in the canal.

the most characteristic canal-side mammal. A less welcome mammal is the brown rat that lives in holes in the banks or under a hedge.

Where a luxuriant growth of marginal plants merges with grasses, willowherb, brambles, sedges, nettles, sallows and hawthorn the ecology starts to resemble that of a scrub-covered hillside with, in season, breeding birds such as the whitethroat and lesser whitethroat, yellowhammer and linnet, together with the two most typical birds of the canal-side, the reed bunting and the sedge warbler.

Maintaining a working canal The changes that occur in a working canal are slight. The ecological balance is maintained by the very fact that the canal is being used. Navigation channels must be kept open, banks maintained in sound condition, depth kept constant by periodic dredging and random seedlings of trees and shrubs cut to ensure that the towpath does not become an impenetrable forest. Therefore a walk along a canal today is likely to produce the same bird species, mammals and flowers that you would have seen a century ago. The only things missing are the horse-drawn boats.

Disused canals In a canal where use and management ceased long ago, a fascinating type of wetland habitat has the chance to develop. The water level is no longer maintained and falls slowly, the vegetation growing at the margins spreads out from the banks, the central navigation channel becomes shallower and narrower, and new plant species start to gain a foothold. Uncommon plants such as frog-bit and water soldier may occur in great profusion, with great patches of yellow loosestrife on the canal banks.

Yet, the profusion of wildlife is just a transitional phase. Left to itself, the disused canal transforms completely. Decaying plant debris gradually replaces the water as species such as the common reed extend their dominance, trees such as willows, poplars and alders become established and the true aquatic plants and the insects associated with them disappear. There is eventually little to remind you that this was once a busy working waterway.

THE FRESHWATER RIVER THAMES

The non-tidal Thames, from its much-disputed source near Cirencester to Teddington Lock, meanders among water meadows, woodland and grassland, cutting a deep gap through the chalk at Goring. The river enriches the landscapes and habitats it passes through.

The source of the River Thames is a spring in a clump of trees at Trewsbury Mead, about 5.6km (3½ miles) south-west of Cirencester in Gloucestershire. It can be reached by walking north from the old Roman road, the Fosse Way – now the A433 – although it is on private land and there is no public footpath. The spring is marked on the Ordnance Survey map as 'Thames Head: the source of the River Thames' – and the Thames Water Authority recognise this site as the official source. However, the source of the Thames has often been disputed. Some claim that it is at Seven Springs, about 4.8km (3 miles) south of Cheltenham.

In summer there is seldom any water flowing from the spring; the Thames is a winter-bourne here and flows on the surface only during the winter months. In summer the spring and watercourse may be completely dry down to Somerford Keynes or further. Except for the dry saucer of stones marking the spring, the source of the River Thames is ploughed up and sown with rye grass – the modern agricultural replacement for the traditional meadow or pasture which once existed here and was cut and grazed with great regularity every year.

Rising in the Cotswolds, the Thames flows eastwards towards Oxford, starting as a clear, rippling, infant stream, then running swift and shallow with no weirs to impede its progress between Cricklade and St John's Lock

Above: The Thames in tranquil mood – the view upstream from near Pangbourne. The wooded hillside on the right belongs to Oxfordshire, the bank on the left to Berkshire.

Left: The water meadows bordering the Thames are famous for their snake's head fritillaries.

Opposite page: In June whole fields by the side of the Thames are carpeted with buttercups and other meadow flowers.

Typically, they are flooded in winter by the increased amount of water carried by the river and so are suitable for grazing or cutting only in the summer. Once drained, then the grazing or cutting season can be extended and the diversity of flora reduced.

Grassland flowers A typical plant species in the meadows and pastures of the upper Thames is the beautiful snake's-head fritillary. It has declined dramatically as a result of drainage and increased grazing, and picking, particularly for sale, has also contributed. Efforts by the Berkshire, Buckinghamshire and Oxfordshire Naturalists' Trust and the Gloucestershire Trust for Nature Conservation have resulted in the traditional management of some meadows and pastures being retained for the conservation of this and other species in these wetland areas.

In Britain orchids are rare in any plant community, but in the wet grasslands bordering the Thames the early marsh orchid is a fairly common species, green winged orchid is local (but abundant where it does occur), and burnt orchid is uncommon. Adder's-tongue fern is common in some riverside grasslands.

Rare plant species are not the only attraction of riverside grasslands. Many common species contribute to their beauty. Whole fields coloured golden with dandelions or buttercups are typical in June, and among these dominant flowers are patches of common species which provide exciting splashes of colour. These include the brilliant pink ragged robin, meadowsweet, purple-red great burnet and shuddering quaking grass.

Chalk woodland The Thames rises in the limestone rocks of the Cotswolds and flows

(Lechlade). At Oxford it bends sharply to flow southwards to Abingdon and onwards to the Goring Gap where it flows between the thickly wooded chalk Chilterns and Berkshire Downs; the river here becomes broader and slower moving–and also muddier. At Eton and Windsor it flows south-east, but turns at Weybridge to flow east again through London to the sea.

On its way the Thames passes through dramatically changing landscapes – water meadows, agricultural grassland, arable crops, wooded hillsides and built-up urban areas. The flora and fauna of all these different habitats are quite distinct – contributing much to the diversity and richness of this greatest of English rivers.

Riverside grasslands The characteristic habitats bordering the Thames from source to estuary are water meadows and pastures. Meadows cut for hay, and pastures grazed by livestock, have been the traditional type of farming in the Thames Valley for centuries. Although these grasslands are artificial habitats, created by man, they are full of native species of grasses and herbs. If left unmanaged, they would eventually succeed to deciduous woodland, dominated by oak, alder or willow according to the differing environmental conditions.

However, there is not much chance of this today. Most of the grasslands have been destroyed by urbanisation and modern intensive agriculture. Drainage, ploughing, herbicides, fertilisers and agricultural grass seed mixtures have converted these areas into high yielding grasslands or arable crop fields. Trewsbury Mead is no exception.

Nevertheless, a few ancient water meadows and pastures still exist, particularly from the source of the river down to Oxford and just below. These meadows and pastures, situated on the alluvial soils of the Thames flood plain, are some of Britain's richest grasslands in terms of number of species growing on them.

Above: The source of the Thames–Trewsbury Mead near Cirencester. A statue of Old Father Thames once stood here, but it was removed for safety and now stands beside the river at St John's Lock, Lechlade. There is confusion over the name of the stream flowing from Trewsbury. From its source to its junction with its tributary, the Thame, some distance below Oxford, the ancient name for this part of the river is the Isis. It only becomes the Thames after joining the Thame.

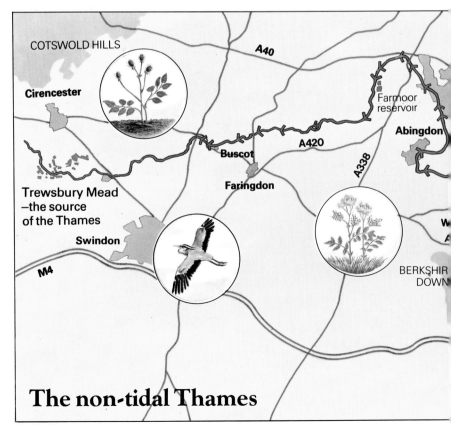

The non-tidal Thames

through deposits of gravels, sands and clays to the Goring Gap where, at some time in the past, it has cut a path through the chalk hills, dividing the Chilterns from the Berkshire Downs. It then flows onwards through more gravels and sands to the estuary.

From Goring to Maidenhead the river passes by steep chalk hills covered with grassland and woodland (mainly beech). Chalk grassland is as rich in species and as colourful as the wet meadows. Beech trees in the woodland cast a dense shade because of the overlapping arrangement of their leaves. The lack of light means that very few plants are able to grow beneath them. However, bird's nest orchid and many toadstools are well-adapted to this dim light. They are saprophytes, digesting the leaf litter for food and so unlike green plants as they require no light.

Wetland birds The extraction of sand and gravel throughout much of the Thames Valley has produced numerous water-filled pits. The construction of reservoirs, such as the ones at Staines and Farmoor (near Oxford) has also increased the area of fresh water. Gravel pits and reservoirs provide valuable feeding, nesting and roosting sites for a variety of birds, and their construction has certainly been responsible for the increase in numbers of the great crested grebe. The newly dug edges of the gravel pits provide breeding sites for the little ringed plover, which first bred in Oxfordshire in 1947, while the vertical cliffs of the pits are often riddled with the nest holes of sand martins. The increase in numbers of Canada geese is probably also related to the proliferation of gravel pits, which they use for breeding.

Above: The arrowhead is a water plant you may well find in the quieter, shallower backwaters of the Thames. It is named for the distinctive shape of its leaves.

Much of the land along the river has been turned over to arable crops — mainly wheat and barley — and the grassland has been sown with a mixture of about five grass species and clovers. The traditional meadows are recognisable by their wealth of wild flower species.

This increase in gravel pits and reservoirs has helped to make up for the loss of marshes and shallow water due to drainage operations. Ducks which favoured these areas now make use of the pits: mallard, tufted duck, wigeon, teal, pochard and shoveler all breed on them, while winter visitors include goldeneye and, more rarely, the long-tailed duck.

All these ducks, and others as well, can be seen regularly on the waters of the Thames Valley – but because of the constant flow of boat traffic from spring onwards through the summer, they prefer to nest away from the busier stretches of the river where there is less chance of disturbance to the eggs and chicks. The gravel pits are, of course, ideal as breed-

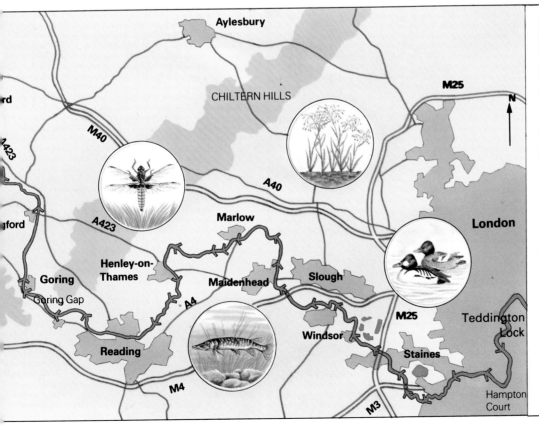

From its source among the limestone Cotswolds, through the flat water meadows and on past the beech woodlands of the chalky Chilterns and Berkshire Downs, the Thames takes a slow, meandering course. There are many locks along its length – and a vast number of boats that negotiate them every year! Anglers, tourists and day-trippers also abound, but there's plenty of wildlife species still to be seen – among them the great burnet, heron, meadowsweet, dragonfly, pike, quaking grass and goldeneye duck shown here.

→ lock
— road
☐ town
= motorway

Above: The Loddon lily is a very rare plant species which is only native along the River Thames and the River Shannon. Its name reflects its association with the River Loddon—a tributary of the Thames. Another name for this attractive species is the 'summer snowflake'. It flowers in April and May and reaches 60cm (2ft) in height.

Left: A solitary heron hunched on a post. Herons and kingfishers are still common along the Thames and its tributaries. Heronries have declined in the Thames Valley, but a thriving one can be seen at Buscot Park near Faringdon.

Opposite page: Among the wading birds, the redshank still breeds in the wet meadows and pastures lining the Thames as do both the curlew and snipe; though numbers of all three species are declining.

ing sites.

Insects During the summer, dragonflies, damselflies, mayflies and caddisflies are frequently seen on the wing along the river banks. They all spend their larval stages underwater. In the upper reaches of the Thames in June scores of damselfly nymphs crawl up the sedges and emerge as brilliant blue adults. In still pools out of the way of the rippling current, whirligig beetles and pond skaters run on the water surface with their specially adapted legs.

Thames fishes The Thames and its tributaries offer an excellent variety of fishing. Evidence of its popularity with anglers is provided by the number of rod licences issued by the Thames Rivers Authority. During the year 1991-2, for example, more than 103,000 annual licences, 28,000 fifteen-day licences and 25,000 junior licences were issued.

From Lechlade to Teddington Lock—the navigable part of the section of the Thames considered here—anglers are likely to catch such fishes as bream, gudgeon, pike, chub, perch, and roach. The faster-flowing stretches of the river harbour barbel, a bottom-dwelling fish, while in a few places the tench—also a bottom-dweller—may be found. Shoals of bleak and dace can be discovered feeding near the surface of the water.

Brown trout are abundant in the Thames above Lechlade, where the river is narrower and faster-flowing, these fishes favouring the limestone and chalk tributaries of the river, and also the weir pools between Lechlade and Teddington.

The National Rivers Authority has done much to clear pollution of the Thames, so the river is much cleaner now than it was 100 years ago. As the water quality continues to improve, we may even see salmon breeding in the Thames. The National Rivers Authority are experimenting with the placing of salmon eggs and fry in some of the tributaries of the Upper Thames and salmon are being caught by anglers—good evidence that these salmon are thriving and an indication of the high quality of water in the Upper Thames catchment area.

The plight of the mute swan

Mute swans, like most birds, eat sand and gravel which lodges in their gizzards and helps to grind up plant food. Unfortunately, swans also pick up lead weights lost by anglers. The soft lead is ground up in the gizzard and then absorbed into the swan's tissues. If enough weights are eaten, the swan will die. Death from lead poisoning is a long, slow process. Swans suffering from lead poisoning have a kink at the base of the neck (right), whereas healthy swans are able to hold their heads in an upright position. The lead affects the nervous and muscular systems so that the swans become lethargic and cease to eat because the gizzard muscles no longer function.

National River Authorities in England and Wales have now banned the use of lead weights by anglers on their waters, and these are being replaced by weights made from non-toxic materials such as stainless steel. However, swans will continue to pick up existing lead weights for some years to come.

THE LONDON THAMES

London's Thames, hemmed in by a desert of stone and concrete, looks murky and lifeless but actually has a wealth of wildlife interest.

If you stand on London Bridge and look over the parapet, the Thames slides by beneath, its water greeny-brown and carrying pieces of driftwood and plastic cups and bags–the flotsam of a busy urban river. The Thames here is a working river, with tugs towing lighters carrying timber, fuel and London's rubbish and, in summer, passenger launches packed with tourists. Running between constricting stone walls for much of its length, with the buildings of London rising all around, the river can look devoid of life.

These days, however, the Thames is not a dead waterway but a living one, full of wildlife

interest. Part of its attraction is that it is always changing. Being tidal, it is sometimes so high that it is within feet of the top of the bankside flood defence walls when fresh water from recent rains on the Berkshire Downs and Chilterns roars downstream to meet the high tides running up from the sea. Nevertheless, only three or four days later the river will be flowing placidly and smoothly again, with extensive gravel and mud banks left on either side of the water.

Moreover, as it meanders along no two reaches are ever quite the same in appearance. Downstream from Teddington to Richmond

Above: The Thames from Richmond Hill in autumn–at this point the river bed is mostly gravelly and the banks are grassy.

Right: Black-headed gulls in winter, dabbling in the river mud at low tide.

Below: A haddock–this species has been caught in the Thames off Barking in recent times. Whiting, too, are now common and even mackerel have been found.

the river has a natural aspect: the banks are grassy, trees grow close to the river's edge and the river bed is gravelly for the most part (but muddy on the bends). Small islands, or aits, break the river's gentle flow and offer sanctuary to aquatic animals which thrive out of the current, at the same time providing shelter for water birds and insects. Below Richmond Bridge the river has an idyllic quality, with Kew Gardens on the south bank and the banks of Syon Park on the north.

Below Kew Bridge the river tends to become busier, with pleasure boats and private craft: through Mortlake, Corney and Chiswick reaches the banks are increasingly man-made, although grass and trees make a natural border to the river itself. Buildings and roads begin to intrude into the background from here downstream. From Wandsworth Bridge the river banks are increasingly artificial and the river bed becomes muddier, with occasional exposed gravel banks.

Just downstream of Barking Creek, at the mouth of the Essex River Roding, the great sewage works of Beckton on the north bank and Cressness on the south pour millions of gallons of highly treated sewage effluent into the river. Here, where the wealth of land allows a wilder landscape, the river wall dominates the view; beyond it the landscape is increasingly flat as the estuary begins.

Thames tides In the London area the Thames is dominated by the sea, the rise of the tides causing the river's water to be ponded back and the fall providing constantly varying glimpses of the edges of the river bed. Twice a day, approximately, the tide falls so that the river water can flow directly to the sea, but just as frequently it is pushed back on the flow of the tide. At the neap tides, when the sea water rises and falls less, the ponding back is less noticeable than an spring tides. Even the spring tides vary in level with the seasons, from the strong movements at the equinoxes to the moderate tides of mid-summer and mid-winter.

The tidal nature of the river has a bearing

Fish fortunes 1957–1993

For anyone wishing to fish in Central London in the early 1950s the prospect would have been extremely poor; the river was so polluted that no fish species at all could live in it. Thirty miles or more of the Thames were lifeless. Pollution control measures have, however, improved things greatly, as our chart shows.

Species	1957	1963	1967	1973	1977	1993
Bass						✓
Bullhead						✓
Carp						✓
Salmon						✓
Stickleback (three spined)					✓	✓
Rainbow trout					✓	✓
Brown/sea trout					✓	✓
Common goby					✓	✓
Sprat					✓	✓
Pike					✓	✓
Perch					✓	✓
Flounder				✓	✓	✓
Smelt				✓	✓	✓
Dace				✓	✓	✓
Bleak				✓	✓	✓
Bream			✓	✓	✓	✓
Roach		✓	✓	✓	✓	✓
Eel	✓	✓	✓	✓	✓	✓

Above: Pochard can be seen around Woolwich in winter.

Below: Tufted duck swim near the Tower of London.

on all aspects of its natural history and its use by man. For one thing, the amount of salt in the water varies with the state of the tide and with upland flow (fresh water coming downstream). This means that animals which favour low salinity water can choose the area in which they prefer to live to give them minimum exposure to adverse conditions.

This is illustrated by the distribution of various species of small oligochaete worms. Near Wandsworth Bridge *Tubifex tubifex* is dominant in numbers, downstream this species and *Limnodrilus hoffmeisteri* are dominant at London Bridge, and down at Woolwich a third species, *T. costatus*, is numerically dominant. These worms are important food for several fish species when the tide is in, and for ducks and waders when it is out.

The tidal nature of the London Thames has caused problems, however. There was always the possibility of flooding when high fresh water flows coincided with spring tides and strong northerly winds over the North Sea–this is why the Thames barrier has been built at Woolwich. Another problem is that, due to its tidal nature, anything put into the water in London does not simply wash downstream but is carried up and down the river at the mercy of the tides.

River pollution The tidal nature of the river was ultimately the cause of pollution in the River Thames, which resulted in it becoming a lifeless river over much of its length for many months of the year in the 19th century and again in the 1940s and 1950s. As London grew, most of the waste from its industry, slaughterhouses and fish market, and sewage found its way directly or indirectly into the Thames. Organic matter of this kind decays by bacterial action in water, using oxygen dissolved in the water in the process. As a result there is less oxygen available for fishes and invertebrate animals. In the 1860s the river was practically lifeless and it smelled vile as well.

As a result of public concern about the state of the river, mainly as a threat to health, a major new drainage system was built which took most of the sewage eastwards of London, to Beckton and Crossness where, after treatment, it was discharged into the river. However, the river is tidal there, too, and the result was mostly just to move the worst pollution downstream from central London to the vicinity of Barking and Woolwich.

With short-lived improvements, as different methods of treatment were tried, the Thames was in a poor state during the 1920s and 1930s, but by the 1940s and 1950s there were some years when as much as 48km (30 miles) of river were absolutely lifeless. The concern for public health led the London County Council and the Port of London Authority to initiate a huge rebuilding scheme

Above: A view over the Thames to the barrier. East of the City the once-busy docks became wastelands, providing temporary refuge for wildlife and for plants such as the rosebay willowherb growing here. Now the wastelands are being reclaimed.

The London Thames (taken here to be the section from Teddington Lock to a point just east of Barking) is a tidal river, constricted by stone or concrete walls and lined for much of its route with industry, sewage works, power stations and docks. Among the wild creatures found in, on or near the river are dace (**1**), teal (**2**), pintail (**3**), salmon (**4**), redshank (**5**), mackerel (**6**) and flounder (**7**). The migratory species are among the most exciting fishes in the London Thames. Between 1979 and 1982 nearly 200,000 young salmon were put into suitable tributaries. Many of them travelled downstream to the sea and returned, for the first time in any numbers, in autumn 1982, when over 100 salmon were captured, mostly at Molesey Weir.

at the major sewage works, improving the treatment of sewage while at the same time closing many small, rather inefficient works in the London area, and to control strictly the discharge of polluting substances into the river.

In the late 1950s freshwater fishes could be found in the tideway only between Teddington and Hammersmith Bridge, while sea fishes occurred upstream only as far as Gravesend. There were no fish for the middle 48km (30 miles) of the river, and this also applied to many of the aquatic invertebrates, particularly the crustaceans and insect larvae. Few birds could be seen on the river then, except for foraging gulls, occasional mallard and mute swans.

By the mid 1960s the new pollution control measures and the better treatment given to the sewage had caused the condition of the water to improve. More oxygen was present in the water and fish began to reappear.

Thames fishes Downstream, sea fishes such as tadpole-fish, sand eels, gobies, pipefish and smelt were caught between 1964 and 1966, while upstream roach were captured near Wandsworth Bridge in regions where no fish could be found a decade before. Then for the next 20 years more and more fishes appeared, and in greater numbers, until by 1993, 113 different species had been captured in the tidal river from Teddington to Southend.

Upstream of London freshwater fishes dominate. In the 1970s roach and bleak were most abundant, but dace, bream, carp, tench, pike and perch all occurred frequently. More recently roach seem to have become less common and dace and bleak dominate the upstream reaches. Rainbow and brown trout also occur.

Downstream of London sea fishes are dominant, but their numbers fluctuate with the seasons, and with the spawning success of the species in the North Sea. At various times young sprats and herring have been immensely abundant, some coming up as far as London Bridge. Flatfishes – plaice, dab, flounder, brill and sole – have all been caught in numbers downstream of London. Modest, but nonetheless exciting, migratory fishes like smelt, flounder and elvers are now common all the way up the Thames in London – and salmon too, more than 280 have been caught in some recent years.

Invertebrate life Perhaps because of the numbers of birds and fishes feeding on them, the mud-dwelling invertebrate population has changed considerably over the years, particularly the tubifex worms which occurred at one time in such abundance that the mud in places looked blood-red. Small shore-dwelling crustaceans and water snails are also common and form suitable food for both birds and fishes. It seems now that animal types and numbers in the London Thames are becoming stabilised and the river has recovered well from the years of pollution.

Below: A mallard duck with her family of ducklings. That these, and many other, birds can survive on the Thames today is evidence of how successful the cleaning-up operation has been. The Thames was the first river in this country to be treated in this way and it formed an example to the world. The Thames is London's biggest open space and it is also probably London's richest wildlife habitat. For instance, common terns can now be seen on the river, especially downstream from Woolwich, and it is not an unusual thing to see two or three of these birds diving for small fishes, probably young sprats but possibly gobies too.

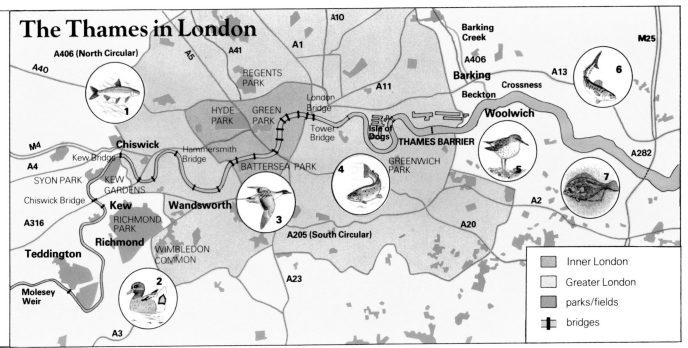

The Thames in London

THE THAMES ESTUARY

The Thames Estuary is an area of paradoxes – great parts of Thames side are taken up with industrial complexes and housing estates, yet close by are marshes and wide open spaces rich in wildlife.

The mouth of the River Thames is a typical trumpet-shaped estuary, widening the further it gets towards the sea. Its limits are difficult to define, particularly towards the sea, for the accidents of history – and prehistory – have produced a series of boundaries. Historically,

the most interesting boundary is marked by two 'stones': the Crowstone on the Essex shore between Leigh-on-Sea and Southend-on-Sea, and the London Stone, near Yantlet Creek on the Isle of Grain on the Kent shore. This Crowstone-Yantlet line, at a region where the river is nearly four miles wide at high tide – but only half that at low tide due to the enormous sand banks on either shore – is the ancient seaward limit of the jurisdiction of the Corporation of the City of London which, from medieval times through to the last century, controlled shipping, fisheries and public health on the river.

The ancient Thames However, the prehistory of the river had a more profound effect on the wildlife of the estuary. Up to and shortly after the last Ice Age the Thames was a tributary of the River Rhine and flowed north-eastwards across the bed of the southern North Sea, presumably to flow into the deeper northern North Sea. At this time the southern parts of the North Sea were covered

Above: Moored boats on the Medway. The River Medway flows into the Thames Estuary at Sheerness.

Right: Common oysters. In the deeper waters of the Estuary, large numbers of oysters occur. Many of the oysters supplied to the London markets come from here – especially the Pyefleet oysters from the Colne mouth. The Essex oyster beds were the site of the accidental introduction of the slipper limpet, a North American mollusc which, by its sheer numbers, drowns out and competes with the more valuable oysters. It is now found along most of the coast of England.

with wooded marshy ground interspersed with river channels and lakes. However, in the immediately post-glacial period the impounded water of the 'North Sea' lake broke through the Straits of Dover and Britain became an island. The Thames, no longer a tributary of the Rhine, became isolated and, due to rising sea levels and general depression of the land in this part of England, continued to diminish in size.

Thus the estuaries of the River Medway on the Kent coast, and the Crouch, Blackwater and Colne on the Essex coast–which today look like river estuaries in their own right–are no more than feeders of the greater Thames estuary, which can be said to extend from the North Foreland on the Kent coast to Harwich at the northern tip of Essex, and encompass the entire Essex coast.

A glimpse of green Downstream of the centre of London, and past the great sewage treatment plants of Beckton and Crossness, open spaces along the river show glimpses of green beyond the river wall around Dartford, near Cliffe, and on the Essex coast below Tilbury. Apart from the industrial horror of Thames Haven and western Canvey Island, and the Medway at the Isle of Grain, the estuary presents a natural aspect, with fields and even marshes, and sand and mud flats on the river side of the sea wall. The first glimpses of sea marsh and saltings can be seen at Canvey Point and on the Leigh Marshes. The Medway flows into the estuary at Sheerness, and from thence on the Kent shore the coast runs uninterrupted, with sand and shingle shores and the huge expanse of the Kentish flats to seaward, none of it deeper than 6m (20ft) until Margate and the North Foreland are reached.

The Essex coastline is as convoluted as the Kentish is smooth. From Southend-on-Sea the shoreline turns north-eastwards along the

Above: Great flights of dunlin can be seen wheeling in the skies over the Thames' estuarine saltmarshes. They come to probe the mud for invertebrates and are often accompanied by turnstones, ringed plovers and oystercatchers.

Right: Common sea lavender –found in abundance on the Thames saltmarshes. It can sometimes be seen covered with a fine web and a mass of ground lackey moth caterpillars. These caterpillars live in a community on the top of the plant and are apparently immune to occasional immersion in salt water from the sea.

secretive Maplin Sands and Foulness, with the tiny River Crouch forming its northern border. It then turns northwards along the Dengie Flats to Bradwell, with its nature reserve and dominating nuclear power station. The River Blackwater joins the Thames estuary here as, too, on the north side of Mersea Island, does the Colne. From thence at first eastwards and then curving north, comes the familiar Essex coast of Colne Point, Clacton, Walton-on-the-Naze, the Naze itself and the little-known Walton backwaters with their secret and private islands, and finally Harwich on the distant horizon.

Much of the land shows the hand of man. Even the farmlands have to be protected by massive sea walls, for this part of England is slowly and steadily sinking into the sea, and the fields must be drained by ditches and channels. Too many parts of the coastline have produced ugly rashes of caravans and chalets for summer use, and the quiet waters attract sailcraft and offensive powered boats.

The sewage sludge from London is dumped in the mouth of the Thames, luckily to be swept north-eastwards into the North Sea. The strong currents of the outer estuary have meant that for a century now London's sewage sludge has been jettisoned into either the Black Deep or the Barrow Deep–which lie seaward of Foulness–without (so far as can be detected) any harm to the environment. Yet, despite all this, there is abundant wildlife.

The saltings Many of the wilder borders of the estuary are fringed with saltings: areas of marshland flooded by the sea at high spring tides in spring and autumn but dry, or comparatively so, the rest of the year. Yet twice a day the tide comes quietly sneaking in to fill the deep muddy channels that snake through the saltings, cutting them a little deeper here, depositing jetsam there, and bringing brief activity to the animals living in the mud.

On an early summer's day the saltings are places of incredible beauty and calm. Fine sea meadow grass forms the main vegetation while on the higher saltings sea lavender grows in dense clumps of purple-blue flowers. Everywhere on these saltings the pale green spikes of marsh samphire stand erect, the older plants tough and branched, the young tips and shoots tender and succulent.

The rewards of these sea marshes can be great. Redshank and curlew are common on the edges of the saltings, great flights of dunlin wheel away against the grey sky, and turnstones probe the mud, together with ringed plover and oystercatcher. The evening flights of duck attract wildfowlers on the quieter marshes, and great flights of Brent geese come in from the sea to graze warily on the newly sprouted corn in the fields, not infrequently leaving huge barren patches. Almost anywhere in the estuary shelducks can be seen in winter, although in the summer they can be

Above: The saltmarshes can be frightening on a winter's day when the wind drives in uninterrupted from the North Sea and a cold light makes each dark channel a menacing chasm. And, indeed, they can be dangerous. It is too easy for inexperienced people to find themselves cut off from the sea wall by a water-filled maze of deep channels, and the knowledge that you could soon be knee-deep in water does not contribute to peace of mind.

Left: Wildlife in an industrial wasteland–mute swans and common reeds on a pit near West Thurrock Power Station.

Where fresh water meets salt
The amount of salt in the water varies with the state of the tide and with upland flow (fresh water coming downstream). As fresh water is less dense than sea water, the fresh water forms a wedge and flows downstream on the surface of the sea water which, also wedge-like, flows upstream. But the course of the tideway is so irregular, with 25 sharp bends between Teddington and the sea, that fresh and sea water become thoroughly mixed and the salt content gradually increases towards the sea.

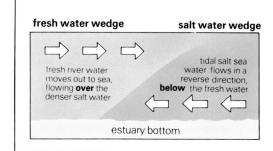

fresh water wedge   salt water wedge

fresh river water moves out to sea, flowing **over** the denser salt water

tidal salt sea water flows in a reverse direction, **below** the fresh water

estuary bottom

found only on the quieter stretches. However, many of these waders and ducks make their way up the Thames in winter and can be seen on the foreshore resting, or feeding on hidden invertebrates in the mud against a backdrop of such electricity generating stations as West Thurrock and Littlebrook.

Other estuary animals The Thames is a typical estuary as regards the animals found there, with a relatively impoverished fauna in terms of numbers of species. However, these species are often represented by millions of specimens. Fishes are a good example of this. There are probably only 50 kinds which are at all common, and fewer than 20 are abundant, yet it is possible to catch thousands in a short space of time using appropriate methods. Young herring are abundant in winter and early spring in the Thames mouth, as are sprats, which tend to be present and leave earlier than the herring. The whitebait (young) stage of both forms is important food for terns, both during the breeding season on Foulness and near Colne Point, and throughout the river estuary during the rest of the year.

The sand banks (often rather muddy, depending on local currents), are important nursery grounds for flatfishes; huge numbers of dabs, plaice and sole congregate for the rich feeding in the sheltered tidal waters of Maplin Sands and the Blyth Sands. Flounders, too, are abundant all the way up to Teddington. Young whiting also flourish, especially in late autumn and early winter. Smelt and eels are abundant as well, being typically estuarine fishes. Larger fishes appear in good numbers seasonally; cod, for instance, migrate southwards in winter and become common all along the outer estuary. In the warmer seasons they are replaced by large numbers of bass, grey mullet, thornback rays and two sharks –the smooth hound and the tope.

The thornback and the less common sting ray and the two sharks are attracted by the rich invertebrate fauna of the estuary. Brown shrimps abound in huge numbers on the sandy and muddy bottoms, pink shrimps and prawns are found in numbers in the deeper water where the bottom is hard, and the sand banks all along the estuary abound with cockles, which form the major food of oystercatchers and sting rays. But perhaps the most typical animal of the estuary is the shore crab. Because of its abundance it is an important food resource for many of the Thames estuary fishes and birds, and it is an equally important predator on smaller animals, as well as being a scavenger.

Above: A view of the Blackwater Estuary, which joins the Thames estuary at Bradwell. The River Blackwater has its own distinctive race of herrings, characterised by their small size and the number of vertebrae and scales (these last detectable only from statistical analysis of large samples). They spawn in April on the Colne Bar, shedding their eggs on the clear gravel in this area. In their first year they form schools in the Colne and Blackwater mouths.

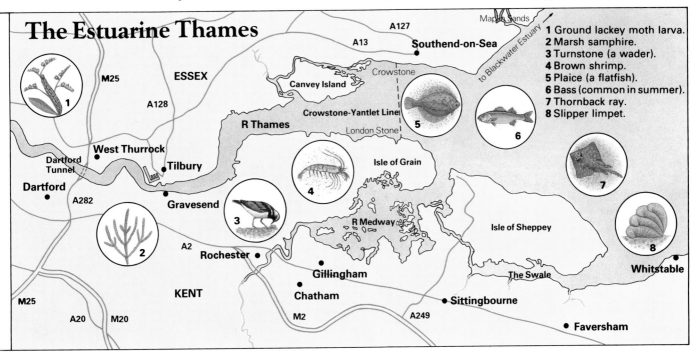

The Estuarine Thames

1 Ground lackey moth larva.
2 Marsh samphire.
3 Turnstone (a wader).
4 Brown shrimp.
5 Plaice (a flatfish).
6 Bass (common in summer).
7 Thornback ray.
8 Slipper limpet.

WATERFALL WILDLIFE

Although Britain's rivers have no giant waterfalls, they are graced by thousands of small ones, their attractions reflected in such names as High Force and Grey Mare's Tail.

Most people, when they think of waterfalls in Britain, consider only those in the Highlands of Scotland, the Lake District and the Snowdonian mountains. However, many smaller, equally spectacular, falls are found in the river valleys of Devon, the mountains of South Wales and in the Pennines.

Waterfall formation Most of our waterfalls have been formed by one of three main processes. First, weaknesses in parent bedrock in the uplands, developed during volcanic and other earth movements, have resulted in upwellings and slippage along fault lines. Where this has happened in river valleys, waterfalls frequently occur. Such falls often have sheer rock faces with turbulent,

Below: The waterfall at Inversnaid, Loch Lomond. The constant humidity created by the falling water gives life to a host of mosses and liverworts in the fall itself, and encourages flowering plants among the surrounding rocks. Waterfalls not only have immense beauty; they can also mirror contrasting moods. The rippling, sun-flecked white water on a summer's day can suddenly change to an all-embracing and terrifying spate of tumbling brown water.

swirling pools at their bases.

Secondly, when the glaciers gouged out their wide U-shaped valleys they left hanging, almost in suspension, feeder streams from the highland glens and gullies. Such streams now often cascade down the steep valley sides, leaping over rock faces and boulder-strewn slopes. Thirdly, some waterfalls, most notably those underground, have resulted from water actually dissolving rock.

Contrasting habitat A waterfall is a habitat of great contrasts. At any moment a gentle trickle can turn into a torrent of tremendous physical force which has the power to remove almost anything in its path. On the other hand, within most waterfalls there are cracks and overhangs cushioned from rapid environmental change. Crevices within waterfalls may be totally enclosed, with niches that receive light and constant humidity, but are protected from any sudden diurnal and seasonal temperature fluctuations by the water passing overhead.

The other major feature of waterfalls, influencing all the plants and animals in their vicinity, is the constantly high humidity they create. In wide valleys, only the crevices and gullies immediately adjacent to the falls are always bathed in a fine spray of water, but in narrow, steep-sided valleys the picture is very different. The moisture content ranges from dripping surfaces near the falls, through the moisture-laden atmosphere of adjacent wood-

Above: The dipper is a specialised bird of rapids and waterfalls. It even swallows stones to increase its body weight so that it can walk and feed underwater in torrent conditions. Birds such as the dipper, the grey wagtail, pied wagtail, goosander, wren and ring ouzel may nest within crevices adjacent to waterfalls.

Right: The greater wood rush is one species able to colonize the banks of falls in low-lying areas.

Below: Pink purslane may well be found growing beside waterfalls, flourishing in the wet conditions. It can be recognised by its pale pink flowers. One of the more unexpected finds in tiny crevices is the sight of many natural 'bonsai' trees. Miniature sycamore, ash, willow and alder trees can often be seen, their constricted root systems providing enough anchorage to resist floods and support minute gnarled branches.

land, to a damp pervasive mist many hundreds of feet above and downstream of the falls themselves.

The rock faces, which endure tremendous erosive flood forces and have been polished to a smooth surface by such forces, have rich plant communities. No flowering plants can withstand the harsh environment, but some algae, liverworts and mosses thrive.

Two types of algae On waterfalls both felt-forming and filamentous blue-green algae can be found. Genera such as *Chamaesiphon* and *Homoeothrix* form very flat crusts on the surface of rock, the former producing minute spores at the crust/water interface, while the latter have hair-like filaments pointing away from the rocks. Slimy pelts of brown and various hues of green are frequent on the lips of falls. These are produced by *Phormidium*, which is composed of tightly interwoven filaments pressed close to the rock surface to reduce the chance of being washed away during floods.

There are also the tough, leathery filamentous tufts of *Tolypothrix*; this blue-green alga, like many of its relatives, is an important component of many nutrient-poor upland communities because it has the ability to 'fix' atmospheric nitrogen.

Simple blue-greens are not the only algae to occur on waterfalls. *Lemanea*, a filamentous red alga, is common on waterfalls in spring, but during the summer months is replaced by

a variety of filamentous green algae.

Liverworts and mosses Whereas much of the algal growth on waterfalls is somewhat transient, the cushions and colonies of the liverworts and mosses are much more permanent. In the uplands, liverworts dominate, only a few mosses being found. Many small falls are smothered in cushions of *Nardia, Marsupella, Scapania* and *Solenostoma* species, all tightly pressed to rock surfaces to avoid being dislodged during spates. Occasionally the less common *Plectocolea, Cephalozia* and *Hygrobiella* also occur, but these are more common in the broken waters of rapids. In the large rivers in the highlands the only common moss is *Blindia acuta*, a small plant with very tough, pointed leaves which resist the pulling force of torrent flows. *Schistidium agassizii* is another moss of upland waterfalls, but until recently had never been recorded in Britain. A classic species of Scandinavian snow-melt rivers, it has now been found in rapids and waterfalls of rivers in many parts of North Wales, the Lake District, the Pennines and also in Scotland.

Where waterfalls occur downstream on rivers, the typical liverwort-dominated community of the uplands is replaced by a moss-dominated one. Instead of tight, rounded cushions, the mosses form less compact growths which are attached to the rock by holdfasts. Many of the shoots are densely branched with aerodynamically curved tips

Above: The leap of the salmon as it struggles to overcome a waterfall is an extraordinary feat of skill, power and endurance. In general, few animals are particularly associated with waterfalls. For some, such as the migratory salmonids, they may even form a barrier, above which point in a river system they cannot penetrate. Where falls are passable, salmon frequently make use of the tremendous energy they can produce by jumping powerfully from upcurrents in the swirling pools below. Young eels also migrate upstream, but they are assisted by their remarkable ability to wriggle up vertical rock faces.

Opposite page: The Braan Falls in Tayside (Scotland) is an excellent place for watching leaping salmon in autumn.

Left: The bank sides and rock fissures near waterfalls provide an ideal niche for the bright yellow marsh marigold which bursts into flower in the spring.

which serve to deflect water. Typical species are *Scorpidium scorpioides, Hygrohypnum* species, *Fontinalis squamosa* and *Rhyncho-stegium riparioides.* On waterfalls which have base-rich water flowing over them, *Amblystegium fluviatile* and *A. tenax* often replace the other mosses.

Plants of the splash zone Adjacent to waterfalls are dripping rock surfaces and crevices that are frequently smothered in vegetation. Again, it is the mosses that dominate, but some lichens, ferns and flowering plants are also conspicuous.

In upland waterfalls between the dry flow level and the flood level, the dark, tough, leathery lichen *Dermatocarpa fluviatilis* is very common, and occasionally the much rarer and lighter *D. leptophyllum* may also be found. Where the rock is base-poor, *Racomitrium aciculare* may be very common, growing among the lichen colonies. Where the rock is more base-rich, *Schistidium alpicola* and *Cinclidotus fontinaloides* replace *Racomitrium.* Frequently, too, many of the same species that occur in the permanently inundated parts of the fall also occur in the zone that alternates between being desiccated and being abraded by flood water. Such plants often show clear signs of living in this harsh environment, being tough and black.

Plants within the more protected environment of crevices and cracks in the splash zone are very different. They are usually a luxuriant

green in colour, and frequently there is so much competition for the prime sites that one plant species grows on top of another. It is a common sight to see cushions of *Dichodontium, Dicranella, Rhizomnium* and other mosses with other plants rooted into them. The hard fern is one example of a common fern found in such a habitat, and the brittle fern is a rare plant which requires just such a moisture-laden habitat. In the highlands the easily confused bog pimpernel and the alpine willowherb may be seen straggling over moss cushions. Both have delicate pink flowers, but the latter's elongated fruits easily distinguish it in late summer. Procumbent forms of the starwort *Callitriche stagnalis* are also commonly seen growing as part of the splash zone community.

On more low-lying waterfalls, adjacent rock fissures invariably have rich higher plant communities. Unshaded spots may burst into bloom with the bright yellow spring flowers of marsh marigold and globe flower, followed later by the delicate blue of water forget-me-not, pink purslane, white scurvy grass and the multi-colours of monkey flower. Among these tiny wild gardens can also be found rushes, sedges and grasses.

Waterfall insects Few insects are specialised for living in waterfalls, although some are ideally suited to take advantage of the lack of competition from others. Most are there because they need to be constantly bathed in oxygen-rich water. Most frequently, insects take advantage of the shelter offered by the leeward edge of any submerged obstructions. In the most rapid falls it is unlikely that any insects other than those which are permanently attached to rock surfaces will be encountered. The commonest examples are minute case-living caddis larvae and the larvae of blood-sucking blackflies, both of which may gather on exposed rocks.

In cascading falls where the rock faces are not always sheer, it is possible to find mobile insect larvae which are specially streamlined or flattened to avoid being washed away. The best examples are some mayflies and stoneflies, with species of *Ecdyonurus, Rhithrogena* and *Perlodes* often present. Some mayflies have even developed hooks to cling to rough rocks. More varied populations occur within the protection of the liverwort and moss communities and in the shelter offered by minute nooks and crannies.

Above: A stonefly on a mossy stone; the aquatic larvae of this insect find refuge in the shelter of mossy nooks and crannies and take advantage of the extremely oxygen-rich water of waterfalls.

Right: Another species to find a refuge among the moss of waterfalls is the alpine willowherb, a plant which grows in mountain districts and bears elongated fruits in late summer.

Waterfall plant-life

Above, left: **1** *Blechnum spicant* fern; **2** liverworts *Scapania undulata, Solenostoma triste*; **3** mosses *Rhynchostegium riparioides, Amblystegium fluviatile*; **4** liverwort *Pellia epiphylla*; **5** common sedge, globe flower, water forget-me-not; **6** moss *Dicranella palustris*; **7** filamentous algae.

Two types of waterfall

1 Where the rock beds are horizontal, soft rock underlying a layer of harder rock is cut back by the water and a deep pool erodes at the base of the fall. The soft rock behind the fall may form ledges suitable for plants.
2 In a vertical or steeply inclined stratum, hard rock prevents the progress of upstream erosion. The soft rock at the base of the fall erodes slowly, producing a shallower pool. There are few ledges for plant life.

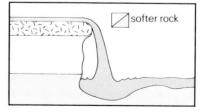

Above: Animal life in the turbulent pool is almost minimal.

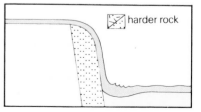

Above: The calmer, shallower pool may harbour water insect larvae.

Waterfalls of Britain and Ireland

1 Eas-Coul-Aulin, Unapool; 2 Falls of Measach, Corrieshallock Gorge, Ullapool; 3 Fall of Foyers, Loch Ness; 4 Dog Falls, Glenaffric; 5 Allt Coire nam Bruardaran, Isle of Skye; 6 Braan Falls, Tayside; 7 Inversnaid Falls, Loch Lomond; 8 Alva Falls, Alva; 9 Bonnington Linn, Lanark; 10 Corra Linn, Lanark; 11 Grey Mare's Tail, Moffat; 12 Skelwith Force, Ambleside; 13 Lodore Falls, Derwentwater; 14 Aira Force, Ullswater; 15 Dungeon Ghyll Force, Grasmere; 16 Scale Force, Buttermere; 17 Low Force, Teesdale; 18 High Force, Teesdale; 19 Cauldron Snout, Teesdale; 20 Catrake Force, Keld; 21 Kisdon Force, Keld; 22 Thornton Force, Ingleton; 23 Pecca Falls, Ingleton; 24 Hardraw Force, Wensleydale; 25 Aysgarth Falls, Wensleydale; 26 Mallyan Spout, Goathland; 27 Kinder Down Falls, Hayfield; 28 Swallow Falls, Capel Curig; 29 Aber Falls, Llanfairfechan; 30 Conway Falls, Betws-y-Coed; 31 Pystyll y Rhaeadr, Clwyd; 32 Mawdach Fall, Dolgellau; 33 Dolgoch Fall, Tywyn; 34 Rheidol Falls, Devil's Bridge, Dyfed; 35 Sgwd Ddwli, Powys; 36 Scwd yr Eira, Glyn Neath; 37 Becky Falls, Bovey Tracey; 38 Lyd Falls, Dartmoor; 39 Rocky Valley Falls, Tintagel; 40 Glencar Falls, Glencar, Co Leitrim; 41 Powerscourt Falls, Dublin; 42 Torc Falls, Killarney, Kerry.

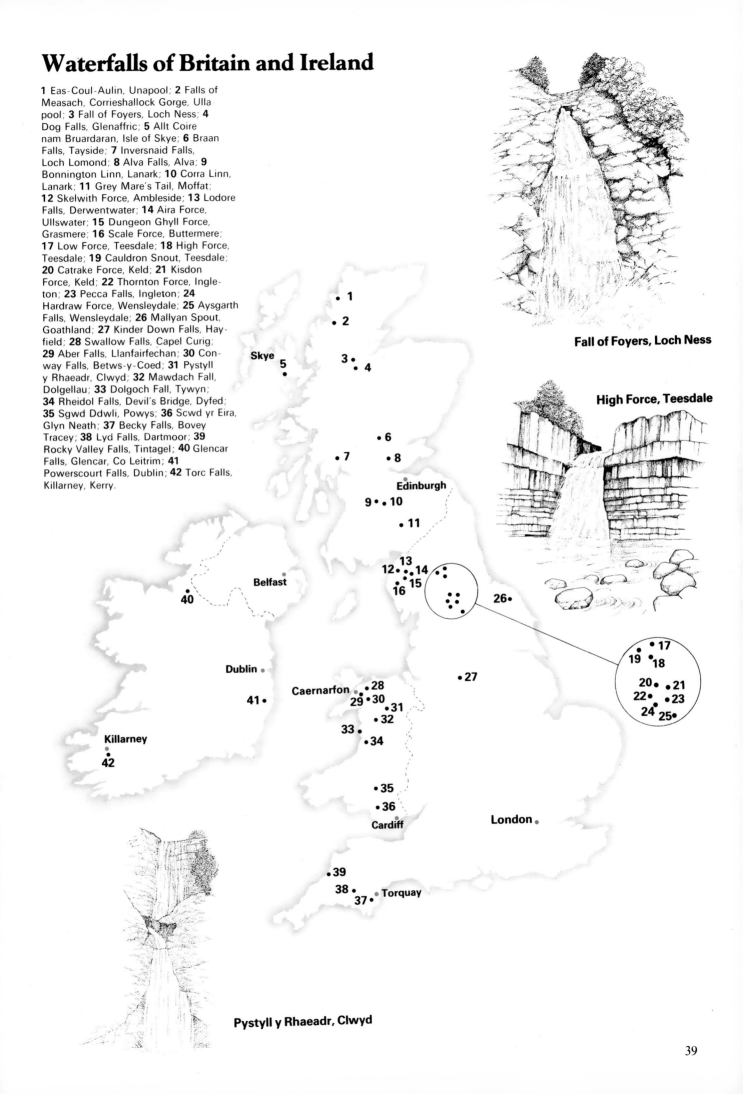

Fall of Foyers, Loch Ness

High Force, Teesdale

Pystyll y Rhaeadr, Clwyd

39

TURBULENT WHIRLPOOLS

Although popularly associated with deep, sinister water swirling below a waterfall, whirlpools may occur in every river in Britain. As such they can be very varied indeed.

Whirlpools may result from a wide variety of natural processes; hard rock, soft rock and natural obstructions are all important in their creation. Man-made obstacles also contribute to their formation, many being the unwitting result of the building of dams and weirs. These and a host of other obstructions to the flow of a river create whirlpools.

To river wildlife, whirlpools are an invigorating extension of an already diverse habitat range. In the uplands they may be the only place where deep water occurs, and provided the pool is large enough it is likely to have many micro-habitats within it. Deposits of fine gravel are common among large boulders while abrasive water forces may act on some

surfaces which themselves produce sheltered alcoves. In the lowlands whirlpools may be the exact opposite of their upland counterparts–the only section within a sluggish river in which visible movement is always evident. Otherwise uniform stretches, which have just sand, silt or clay on the bottom, may fleetingly change to coarser gravels. The water itself may become freshened by the addition of oxygen as it tumbles over obstacles or flows in shallow riffles. In chalk streams, which are characteristically faster-flowing and with gravel substrates, the whirlpool represents the slower, deeper, siltier habitat of the system.

Plants of upland whirlpools Plants in these areas are often highly adapted to a particular

Above: An example of a whirlpool created by sudden constriction along the river banks.

Below: *Calopteryx virgo*, one of several splendid damselflies associated with lowland whirlpools. Other typical species include *Calopteryx splendens* and the white-legged damselfly. Upland whirlpools support the large red damselfly (which can be a common sight as it flits gently between the emergent bank flora), and the large golden-ringed dragonfly.

Whirlpool formation

In upland areas, where the slope of a river can often be steep and varied, whirlpools ranging from as small as 2m (about 6ft) to over 20m (65ft) across are common. Whirlpools of this type are principally formed by rivers descending rapidly in shallow, narrow and sharply inclined channels which suddenly reach deeper sections and are then abruptly 'brought up short' by solid rock obstacles. This results in whirlpools which have great turbulence at their downstream end; this produces dangerous undercurrents. Shown here is a cross-section of this form of whirlpool.

Upland whirlpool

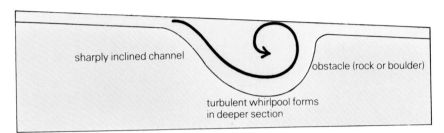

Waterfall whirlpool

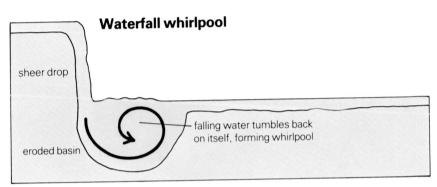

Invariably the downstream section of a waterfall will contain a whirlpool–this is obvious from the very nature of the lie of the land. The whirlpool may be formed by the sheer force of water coming down over the fall at high speed, which makes the water tumble back on itself, or it may be created because the course of the river is channelled to one side or the other, so that lateral swirling of the water occurs. In some cases the two processes may even occur in the same waterfall. Again, our diagram shows a cross-section of this type of whirlpool.

Obstruction whirlpool

Where solid protuberances are found within a river channel which divert the water into a narrow channel, whirlpools are usually found immediately downstream of the obstruction. The increase in velocity at the downstream end of the obstacle frequently results in the formation of a pool by erosion and this in turn leads to a swirling whirlpool. Such pools may produce large 'mushrooming' currents at the point where the downstream flow of water is checked by the rise in the substrate level at the end of the pool. Shown here is the view from above the pool.

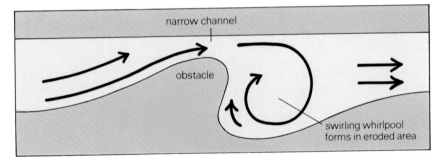

Erosion/deposition whirlpool

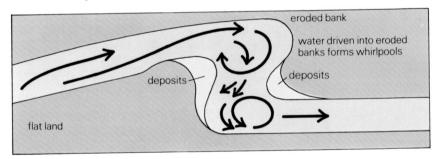

In lowland rivers, the most common type of whirlpool to be found is that caused by the meandering of the channel. This causes erosion on one bank, with corresponding deposition on the opposite bank. In meanders which are particularly sharp and which have therefore eroded a deep pool, the whirlpool is at its most spectacular, for the full force of the water is driven into a sheer, often crumbling soil profile which turns the water full circle. Note that the bank where erosion occurs is steep, while that where deposition occurs is flat. This view is from above.

Tributary interference whirlpool

Where the inflow of a tributary stream is almost the same size as its recipient or main river, a whirlpool is often created. This is particularly true on occasions where the tributary enters the river at a right angle. The whirlpools created here are frequently deep, with unstable, shifting substrates. This is because the discharges of water from the two rivers frequently do not rise and fall consistently; the loose material–gravel, silt and a variety of other debris–on the river bed is consequently pushed first one way and then the other. This view is from above.

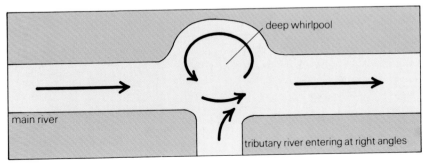

micro-niche within the pool. On rocks which are scoured smooth by the full force of the water, some flattened encrusting algae may brave the elements throughout the year but transient filamentous forms, which are more productive, are absent for much of the year. During spates they are scrubbed clean away. Yet, as soon as the water level and its velocity drops, they start to smother the surface again. Many different species may be found and most of these algae are themselves covered by smaller species of algae growing upon them. This harsh environment therefore fleetingly supports a perfectly adapted community similar to a miniature forest in which an understorey, canopy and epiphytes can be identified.

In cracks between solid rocks, streamlined mosses and tough foliose lichens may be found, some being perfectly adapted to grow totally submerged. Others may be even tougher and able to tolerate both torrential water velocity and periodic desiccation. However, these are relatively less important in whirlpools than they are in other upland river habitats because the whirlpool character-istically does not rise and fall in height so fiercely. Instead, the marginal rocks may be covered by moisture-loving thalloid liver-worts such as *Pellia epiphylla*. On rocks in gently swirling water may be found a whole host of lower plants which are attached firmly by holdfasts. However, where the pool accumulates fine gravels it may be possible to find either the alternate-flowered water milfoil or the intermediate water-starwort.

Animals of upland whirlpools These, of course, depend on the variety of habitats and the plants which these support. Blackfly larvae are characteristic of rocks on the downstream edge of the pools. These areas may be fully exposed to the full force of water, and the blackfly is one of a handful of species which can tolerate such conditions.

The sheltered stones and gravels within the whirlpools are a much richer habitat for invertebrates. In large rivers with pure peaty water the large pearl mussel may be found. The caddis fly *Hydropsyche* is also well-adapted to life among stones or the fronds of

Above: The sheer force of water tumbling over the rock in a waterfall gouges out a hollow at the bottom, in which the water swirls and 'boils' at a furious rate.

Left: Purple loosestrife may often be seen flowering along the banks edging whirlpools.

Below: The marginal rocks and gravels by upland whirlpools may be covered by cushions of mosses, such as the *Dicranella palustris* shown here.

milfoil. Scampering over and under stones, the flattened larvae of mayflies (*Ecdyonurus*) and stoneflies (*Perlodes*) are typical. In small whirlpools in peaty uplands the feathery three-tailed larvae of the large red damselfly or the ferocious-looking golden-ringed dragonfly may be found.

Lowland whirlpool plants These plants differ greatly from their upland counterparts and show marked variation according to the water velocity and nature of the river bed. Where the water is shallow and the substrate is composed of gravels, water crowfoots are usually the dominant plants. On the raceways, weirs, sluices and dams which create whirlpools, two algae are very common: they are the tough and leathery blanket-weed *Cladophora* and the soft pelt-like *Vaucheria*. Both thrive where nutrient-rich water flows rapidly.

Well-vegetated whirlpools are associated with clay rivers in which the substrate is rich but firm enough for plants to anchor. In the deeper parts can be found the yellow water-lily, a robust and resourceful plant. Various species of pondweed may also be seen swirling in the currents.

On the undercutting vertical earth banks which face the whirlpool may be found many colourful annual flowers. If the bank is stable and not eroding, then trees and shrubs often occur and these frequently have exposed root systems which provide habitats for animals.

Animals of lowland whirlpools Among the gravels where the substrate is coarse and the oxygen level is high many mayflies and stone-flies, which are absent from the rest of the river, may be common. Stonefly larvae have a preference for moving water which makes whirlpools an important habitat for them.

The steep banks facing whirlpools may be important as nest sites for such waterside birds as sand martins and kingfishers, the former preferring to nest in soft, open, crumbling banks while the latter like some cover. On flat and reedy margins may be found a wide variety of bird nests, varying from the large mounds built by swans to the precariously balanced nests of coot and moorhen.

The gravels of lowland whirlpools are important for maintaining populations of some fishes. Both the river and brook lampreys occur throughout Britain except in north-west Scotland. However, without well-oxygenated sands and gravels to spawn in they would be absent from lowland Britain too. Whirlpools frequently provide just such suitable habitats—habitats which are also exploited by native populations of brown trout, stone loach and miller's thumb. Many coarse fish are also associated with whirl-pools, which must be deep and well-vegetated. Here the swirling plants provide cover for the invertebrates on which the fish feed, as well as a habitat in which some lay their eggs. Typical species include the pike, tench, rudd, roach and chub.

Above: Whirlpools created by meanders in the river, with erosion one side, deposition on the other.

Right: Amphibious bistort is one of a community of flowering plants that may form on the shallow muddy edges of whirlpools. Other such species include reed sweet-grass and great yellow cress. Purely emergent stands usually contain bur-reed, interspersed with yellow flag, flowering rush and purple loosestrife.

Below: Whirlpools remain the only habitat in some lowland rivers in which the crayfish still occurs. This species requires well-oxygenated, calcareous water and may be found among the fine underwater rootlets of marginal alders and willows.

RIVER SHINGLE BEDS

Shingle is a term normally associated with the coast, yet extensive shingle beds are found in and adjacent to Britain's rivers. They are generally transient and must rank as among our most unstable inland habitats.

Opposite page: Nootka lupins (*Lupinus nootkatensis*) flourishing on extensive shingle banks in the River Dee near Braemar in Scotland. These blue and white flowers are relatives of our common garden cultivated varieties.

Below: Balsam (*Impatiens*) has strong, tenacious roots that enable it to grip firmly on to unstable shingle even when the river is in flood.

In all river systems the presence of shingle is associated with the river's inability to weather its bedrock into fine sands or silt. (Shingle formation results from the ability of a river to erode, transport and then sediment materials.) Thus shingle banks (or bars) in rivers are prevalent in parts of Britain where the bedrock is hard–in Scotland, northern England, Wales and south-west England. In general, shingle bars in the upper reaches of these rivers are haphazardly deposited and composed of small boulders and pebbles, yet in their lower reaches the shingle bars are more stable and composed of much finer gravels. The slope of the river as it races down from the hills and on to the flood plains thus shifts through the easily eroded particles, carrying the lighter, and finer, particles further downstream.

Although shingle bars are characteristic of those rivers which rise in mountains, distinctive examples also occur in the lowlands of Britain. Lowland river shingles are associated with younger rocks which offer some re-sistance to weathering: the Tertiary Sands found in the New Forest, the Greensands of the Weald, the crags of East Anglia, the older and harder sandstones of Devon and South Wales, and the Oolite of the Cotswolds and Northamptonshire Uplands. Even lowland rivers in catchments without hard rocks may also have shingle bars if they flow through coarse drift deposits. These inherently unstable deposits were brought into the area thousands of years ago by retreating glaciers or previous large rivers. Good examples of river shingles derived from glacial gravels are found in the Swale, Ure and Wharfe rivers as they flow into lowland Yorkshire. On the other hand, river gravels are a very common feature of many rivers, and the Wye, Severn and Avon are particularly good examples.

Formation of shingle beds In all cases, the dynamic nature of rivers means that, in some parts of a system, erosion exceeds deposition, whereas in others deposition exceeds erosion. Since these areas may change, the shingle beds which are created are intrinsically unstable

trated by such rivers as the Afon Rheidol and Ystwyth in West Wales. Both rise high on the Cambrian Mountains and make hurried journeys to Cardigan Bay near Aberystwyth. In the flood plain both these small rivers flow in channels up to 400m (1300ft) wide, with the water actually occupying less than 5% of the area; the rest is river shingle.

Most shingles in fast-flowing upland rivers have been derived from a downstream transport of shingle which is all the time being moved both laterally and downstream. Rivers with their catchments entirely in the lowlands frequently do not illustrate large downstream movements of shingle; instead, they frequently erode into banks or throw bed material into small shoals. Shingle beds in lowland rivers are very closely correlated with two features: a meandering channel in which the deflection of currents facilitates natural erosion and deposition zones, and an abundance of gravel derived either from eroded bedrock or from Quarternary deposits. An easy guide to finding lowland rivers with shingle bars is to find out if gravel extraction pits are found within a catchment.

A vivid demonstration of the effect of geology on the ability of a lowland river to form shingle bars is seen in the River Waveney in East Anglia. It rises at less than 30m (100ft) above sea level on clay and flows entirely over this for more than half its fresh-water length until it reaches coarse crags near Hoxne. Above this point gravel beds do not occur in the river, yet below it the river erodes into a flood plain containing gravel and pebbles and shingle banks abound.

Upland shingle The main colonizers of upland shingle are fast-growing, prostrate plants which can exploit bare ground quickly as well as withstand desiccation during low

and short-lived habitats. The processes which lead to their formation thus ultimately lead to their destruction.

In mountainous and upland regions in Britain the shingle beds of rivers are generally composed of large cobbles and pebbles. If no coarse sands and gravels are available, these shingle beds will support no plant species at all because the beds are destroyed and created many times over in a single year as the cobbles roll over one another every time the river rises and falls. The introduction of finer particles provides a firmer base into which the cobbles can become embedded and these shingles can survive all but the largest floods. Eventually, however, they fall victim to the river's immense erosive flood power. As the river subsides to a tranquil trickle it now runs alongside the virgin shingle beds which it created only days before.

As rivers race seawards from their upland sources, they are often contained in their channels by steep-sided valleys, which preclude significant shingle formation. Additional confinement may be provided by tree-lined banks which resist erosion. If, however, a river flows into a U-shaped valley (characteristically created by glaciation), the flat bottom provides ample opportunity for the river to meander and change course. The valley floor therefore does not restrict the river to a discrete channel and shingle banks abound. These river shingles are characteristic of many rivers in Scotland and the Dales rivers of the Pennines.

Flood plain river shingles are a common feature of many fast-flowing rivers as they traverse large expanses of flat land. The rivers, released from the straight-jacket of their youthful confinement to narrow channels, become wayward in behaviour—amply illus-

Above: Coltsfoot in flower on pebbly shingle.

Below: In the last few decades the oystercatcher has made river shingle bars a major nesting habitat—a change which is the key to its recent population increase. The nest normally consists simply of a hollow lined with a few small pebbles or shells. Insects and worms on the shingle are just as good a source of food as the oystercatcher's usual diet of molluscs.

flow and erosion during floods. Although shingle composed of only cobbles and pebbles cannot sustain any plants at all, a number of shingle beds with superficial resemblance to them can support sparse vegetation. Below the surface cobbles of these beds there are fine gravels and coarse sands which provide just sufficient moisture and stability to allow such plants as pink purslane, alpine willowherb and procumbent sandwort to root. Once these establish themselves and add stability to the shingle by trapping detritus, a whole new range of plants can invade.

Prime colonizers of stabilising upland gravels are mosses. Particularly common are such species as *Brachythecium*, *Philonotis*, *Calliergon*, *Bryum* and *Hyocomium*. Despite lacking flowers, these mosses can produce most attractive splashes of colour, the pale yellow-greens of *Philonotis fontana* contrasting with the venous red of *Bryum alpinum*. As the mosses themselves create better water-retention and stability within the shingle, such plants as jointed rush, common sedge, brown sedge, shepherd's cress and eyebright become established. Particular areas of the country have their specialities, with shrubby cinquefoil being a glacial relic left growing only on the coarse shingles of the River Tees near High Force and near Helvellyn in the Lake District. It is a tenaciously rooted shrub which can withstand the full force of torrential water as well as abrasion by the stones carried by the river.

In the middle or lower reaches of upland rivers an immense variety of shingle bed communities can develop. Annuals are the prime colonizers, yet perennials which can hasten their reproductive processes also thrive. Creeping yellow-cress provides a late profusion of yellow flowers where earlier coltsfoot had also flowered. Both thrive because they root very deeply and gain moisture from sands below the shingle.

Among the specialised plants of gravel are to be found many opportunistic species, such as plantains, docks, trefoils, horsetails, bindweeds, knotweeds and many grasses. Yet despite this, different regions or even individual rivers have characteristic plants. In

Above: Extensive shingle beds have been deposited at the confluence of two rivers—the River Severn and the River Dulas. The shingle is partly vegetated—clearly stable enough not to be washed away by every flood. The more vegetation to become established, the firmer the shingle will be.

Left: Pink purslane is one species to look for on river shingle banks. Its pink flowers appear from April to July.

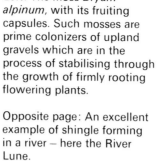

Left: The moss *Bryum alpinum*, with its fruiting capsules. Such mosses are prime colonizers of upland gravels which are in the process of stabilising through the growth of firmly rooting flowering plants.

Opposite page: An excellent example of shingle forming in a river – here the River Lune.

Below: Not many birds frequent shingle bars, but one you may well see is the colourful grey wagtail. It searches among the loose cobbles and gravels for a range of invertebrates.

Bottom: A view of the River Mor, in the Cairngorms. Here boulders and shingle have been brought down by snow melt water.

Scotland the bellflower is a common colonizer of shingle, as are various forms of the pansy *Viola tricolor*. In rivers in northern England the most stable gravels may be colonized by sweet cicely. The gravels of the River Wye, on the other hand, are famed for being the home of wild chives.

In spite of such shingle being only sparsely vegetated, a few birds are often seen feeding upon them. Perhaps the most common is the grey wagtail, which bobs from stone to stone, disturbing insects as it goes. The dipper is by no means confined to feeding in the middle of a rapid stream and frequently it, too, will search for worms and insects among damp shingle.

Lowland shingle The finer particles of the shingle beds deposited by lowland rivers result in greater moisture-retention and a higher nutrient status. The main colonizers are again opportunistic 'weed' species, some of which characteristically also exploit bare ground under cultivation. However, the inhabitants of river shingles must be able to withstand unstable and water-logged soil, as well as flood erosion. Early colonizers are all annuals –marsh yellow-cress, celery-leaved buttercup, knotgrass, red-shank, mallow, toad rush and such grasses as annual meadow grass are typical examples. Some amphibious species, such as the amphibious bistort and great yellow-cress, tend to utilise gravel banks as a step from dry land in their invasion into the water.

Few birds are specialist exploiters of lowland river shingles because they are insufficiently large to provide safe nesting sites during the summer. However, the little ringed plover has, in recent decades, established itself in Britain and utilised areas of loose gravel as nesting sites.

RIVER ISLANDS

The richness of riverine plants and animals is often greatly enhanced by the presence of islands, some of which may be small, temporary and unstable while others may be substantial.

mud, whereas if the rivers flow over less rich soils the islands are composed of compacted sand, silt and gravel. In upland rivers with coarse substrates, and where high rainfall causes frequent erosive flooding, the transient islands are usually very unstable deposits of boulders, cobbles and pebbles.

Temporary islands are usually vegetated only sparsely, and often annuals predominate. When the next large flood comes, many islands are washed away and new bare islands develop elsewhere in the system. However, occasionally the initial sparse vegetation roots sufficiently tenaciously to rebuff the effects of the first flood, and as the water level recedes fine additional deposits are made to the

Above: A temporary island in the River Tweed, with canary grass growing on it.

Below: On rocky islands, where a particular boulder is frequented by birds, their droppings may so enrich the surface that algae and then mosses such as *Funaria* (shown here) may occur. The two birds you are most likely to encounter on and near rock islands are the dipper and the grey wagtail, both of which like to nest in crevices or on ledges.

For the wildlife of a river, one of the major values of islands is their isolation. This can be important for mammals and birds because it may decrease population losses by predation. However, both plants and animals are likely to benefit most if river islands are sufficiently small or isolated enough to make cultivation or development difficult.

Types of islands There are two basic river island types–temporary and permanent ones. Temporary islands can form anywhere within a shallow river where floods cause erosion and there is subsequent deposition of eroded material. In lowland rivers which flow predominantly over rich alluvium or clay, islands are usually composed of soft silt or

River island colonization

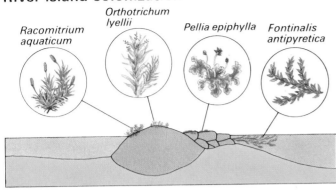

Racomitrium aquaticum *Orthotrichum lyellii* *Pellia epiphylla* *Fontinalis antipyretica*

Primary colonization In initial island formation around a boulder, the submerged surfaces are colonized by mosses.

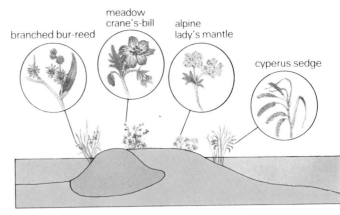

branched bur-reed meadow crane's-bill alpine lady's mantle cyperus sedge

Silt accumulation Once the boulder has begun to accumulate silt, flowering plants can start to colonize the island.

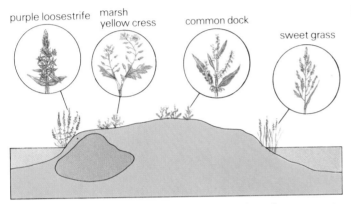

purple loosestrife marsh yellow cress common dock sweet grass

Further growth A wider selection of flowering plants starts to colonize and the island spreads upwards and outwards.

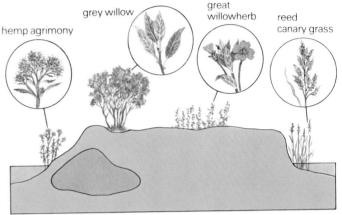

hemp agrimony grey willow great willowherb reed canary grass

The permanent island There is a succession from tall herbs through to scrub and eventually to mature trees.

islands. New species colonize, including perennials with deep or matted roots which start to stabilise the developing islands. As additional species grow, the island becomes more and more stable and eventually willow, sycamore, ash and alder saplings may take root.

All rivers have an exceptionally large flood every ten years or so, which tests a developing island to the full. Most are washed away completely, but if the saplings are rooted deeply, and the branches are sufficiently flexible to bend with the flood water, it is possible that a permanent island will be formed.

For islands or rivers to be regarded as permanent, they must show clear indication of having reached climax woodland vegetation or have been developed for industrial, domestic or agricultural purposes. Such islands are common in lowland rivers but in the uplands they are very rare. However, permanent islands are present in the uplands but these are composed of solid, erosion-resistant rock. These islands are also likely to be small. The smallest are extremely large rounded boulders which have been deposited by glacial movements thousands of years ago or during excessively large floods, whereas the largest rock islands are crags of resistant pavement which the river cannot erode.

Island formation There are many other ways in which an island can be formed, besides its

Right: A bird to look for on islands which are well vegetated, especially if there are alders–the siskin.

Below: The smooth, hard surfaces of rock islands are likely to be colonized only by simple plants. Examples include submerged filamentous algae such as *Ulothrix*, tough, leathery lichens such as *Dermatocarpon*, foliose and thalloid liverworts such as *Nardia* and *Pellia*, and mosses such as the *Racomitrium* shown here.

51

creation from a mere mud deposit in the middle of the river. The formation of rocky islands within rivers is a simple expression of the flow of water taking the easiest route to the sea. If a boulder or solid rock outcrop in a valley bottom is too large to be moved, or too resistant to be eroded, then it is simplest for the river to go round it. If the land on either side of the rock is at approximately the same height the water can be channelled on one side or the other and during floods it is likely that both sides will be utilised. Such floods will make inroads into any superficial deposits and a rocky island will be formed.

Natural earth islands in the flood plains of rivers may also be formed primarily by erosion rather than deposition. The finest examples are on incised meanders where islands are formed during floods. At such times the speed of the water flow may be too great for the meanders to cope with and a new channel is cut which by-passes its previous tortuous path. Such islands may be small or large, but few remain today since agricultural encroachments into the flood plain have necessitated the infilling of the old river course.

Plants of rock islands Extremely large rounded boulders deposited in the middle of a river can rarely support any flowering plant species at all. Although most common in the uplands, boulder islands do occur in the lowlands when a river traverses erosion-

Right: Common persicaria can be found on islands with mature vegetation. A member of the dock family, this plant produces heads of pink flowers from June to as late as October.

Below: An otter in water. Several species of mammals frequent islands, favouring those with large trees and a thick understorey of scrub or reeds. This type of understorey is the preferred haunt of otters, especially if the island is quiet and secluded. If the tall herbs, reeds and scrub are virtually impenetrable to man, the otters may well use the island as a base, from which they can patrol both the main river and its feeder streams. Among other mammals, Daubenton and pipistrelle bats often roost in the hollow trunks of dead or decaying trees. In winter, however, Daubenton bats return to caves and old buildings to hibernate.

resistant rocks. Such islands occur in south-west England, Wales, the Lake District and Scotland. Wherever they appear, however, they all have a great deal in common. In general, their lower submerged surfaces are colonized by such mosses as *Fontinalis* and *Rhynchostegium*, while their exposed surfaces are often bare save for isolated lichens such as *Verrucaria*. Variations do occur, according to the geology of the rock, but the most exaggerated differences are seen where a particular boulder is frequented regularly by birds. Their excrement may so enrich the surface that algae such as *Prasiola* completely turn the surface green. As these in turn provide a thin layer of organic matter on the rock surface, new mosses such as *Funaria, Orthotrichum* or *Tortula* may occur.

On large rocky islands where cracks or hollows accumulate silt, however small the amount, the flora may be extremely rich. In narrow rock fissures showy plants grow in miniature, the restricted root growth reducing vegetative development yet stimulating flower production. The most spectacular species are more typical of neighbouring meadows, with globe flower and meadow crane's-bill providing bright splashes of yellow and violet, while alpine lady's mantle and alpine meadow rue have much subtler shades. The crevices also support grasses, sedges and rushes, alongside which may be found a miniature yellow member of the lily family–bog asphodel. In

rock hollows where most substantial deposits of silt may accumulate, the purple flowers of the common butterwort may be seen among the pink straggling flowers of pink purslane.

Temporary island vegetation The plants associated with the initial colonization of loose material thrown up to create islands by torrential floods is determined by the size of the particles and their ability to retain moisture. Upland islands generally comprise coarse gravels and large cobbles which drain rapidly. As these islands rarely last more than a few weeks, only fast-growing annuals with deep roots survive. In the middle reaches of rivers, where fewer cobbles are deposited and the gravels are finer, opportunist species such as shepherd's purse, common dock, ribwort plantain, redshank and lesser knotweed are quick to colonize. However, these are soon joined by, or replaced by, species more associated with rivers and wetlands in general. Species of sweet-grass, creeping bent, foxtail and canary grass help to stabilise the shingle, while balsam, creeping yellow cress, monkey flower and forget-me-not may produce a profusion of flowering colour. The succession to a mature island normally depends on the establishment of reed canary grass because it has an extensive matted root system which binds the shingle together. If it survives it will spread and trap more silt to enable the island to grow both upwards and outwards. If sufficient silt is deposited, willows will eventually colonize and a permanent island may form.

Islands which start initially from the deposition of fine silt in lowland rivers have a totally different plant assemblage. The early colonizers are again frequently annuals, with toad rush and nodding and trifid bur-marigolds being typical. These may be joined by brooklime and the pink and the blue water-speedwells. On particularly rich mud or in slow-moving rivers, watercress, amphibious bistort, great yellow cress and bittersweet are common. All these species have a combination of most of the prerequisites for success in such transient habitats–they can behave as annuals, grow fast, have a creeping growth or low profile and can root deeply to bind the mud.

Birds of vegetated islands In intensely cultivated river corridors a densely vegetated island within the river channel itself can be an ideal habitat for bird life which has been driven from adjacent land–finches, siskins and warblers all occur here.

Below: Although the building of bridges has destroyed many islands, they are now creating new ones. The islands are formed around mid-stream parapets which act either to collect debris and silt which form islands upstream, or as current deflectors which cause the deposition of fine particles downstream. Despite their isolation, river islands have been exploited by man in many ways–for wool and grain mills and road and rail networks.

Left: Nodding bur-marigold and (right) flowering-rush. For a young mud island to develop beyond the first stages of colonization, when such plants as the bur-marigold appear, it must show a succession to emergent reeds. The commonest is the bur-reed, but this rarely succeeds once the top of the mud is much above the surface of the water. The same is also true for the flowering-rush, but both species are important in the development of islands from mud because they collect debris and silt which raise the level of the deposits. As the mud becomes drier, reed canary grass and reed sweet grass dominate, and alongside these may be found purple loosestrife, hemp agrimony and great willowherb.

53

Trees of river and stream

River and canal landscapes are incalculably enhanced by the variety of shapes, sizes and seasonal hues of their attendant trees – but the human eye is not alone in being attracted to such scenes. The canopy of waterside trees attracts feeding insects and other invertebrates in their hundreds during summer, and when the leaves fall to the water in autumn they become the source of food for a totally different, yet equally hungry, world of minute animal life. The roots and boughs also offer shelter to a host of creatures.

The willow is the foremost species associated with the water's edge, yet its abundance today is related more to the influence of man than to natural development. The most interesting for wildlife is the 'pollard' – created by regular cutting. Ancient pollards, with huge and gruesomely gnarled trunks, often have crowns filled with leaf-litter and are a habitat in themselves.

The alder, unlike the willow, is less plentiful now than in years gone by, but it is Britain's most characteristic waterside tree. Its root system, highly branched, spreads like minute pink fingers into the water, creating a habitat that becomes a haven for aquatic invertebrates.

Many more trees are also found along streams, rivers and canals, yet it is a tragedy that there are not more. Riverside trees paid a high price during river 'improvements' carried out to help agriculture. Once virtually every river below 500m (1650ft) was lined with native trees. In the water-logged sections of carr and fen, alders and willows thrived, and in drier sections trees such as ash, beech and maple, together with shrubby dogwood, spindle or guelder rose, grew where the soil was calcareous, with rowan, silver birch, oaks and Scots pine flourishing alongside rivers flowing over the more acidic rocks. Although millions were removed, not all were lost and vestiges of the natural riverine tree fauna can fortunately still be found.

Left: Summer outline of the Chinese weeping willow, with leaves and catkins of the golden weeping willow. Surprisingly, because they are so widespread, both these species are introductions to Britain, not natives.

CHECKLIST

*This checklist is a guide to the trees you will find near rivers and streams. Although you will not see them all in the same place, you should be able to spot many of them throughout the changing seasons. The species listed in **bold type** are described in detail.*

Alder
Balm of Gilead
Balsam poplar
Black poplar
Chinese weeping willow
Crack willow
Cricket bat willow
Eared willow
Goat willow
Golden weeping willow
Grey willow
Lombardy poplar
Western balsam poplar
White willow

Left: White willows lining the banks of the River Ouse. Willows, more than any others, are the trees associated with freshwater habitats. Willow seeds need to find very moist conditions if they are to germinate successfully, and they must do so quickly, within a few days at most, or they will die.

55

BLACK POPLARS

In early spring the black and the Lombardy poplars bear masses of bright crimson-red catkins that make these water-loving trees an attractive sight in the sun.

The black poplar is usually regarded as being native to Britain. Certainly, it has been growing here since at least the Middle Ages, though it may have been brought over by the Romans from Italy, where it is still very common.

In Britain, the black poplar is confined mostly to central and eastern England. Its typical habitats are wet woodland and the sides of streams, since it likes moist soil, but it does not tolerate stagnant water. The reason for this is that the roots of a black poplar breathe, just as do the parts above ground, so the water in the ground must have oxygen dissolved in it. Therefore the water has to be flowing rather than stagnant.

Catkins in spring Poplars are closely related to willows: both have wind-pollinated catkins as flowers that come out early in the year before the leaves. The catkins of the black poplar mature in March. Male and female catkins are borne on separate trees. Male catkins consist of many tiny flowers, each with a pair of red or purple stamens but with no petals. The female catkins have green flowers, each consisting of a stigma to collect the pollen and an ovary in which the seed is formed.

In May, the female catkins ripen into seeds with long white silky hairs to help their dispersal by the wind. So many fluffy white seeds are released by just one female black poplar tree that the females are never planted in towns – if they were their seeds would collect along pavements and block up gutters.

Triangular leaves The leaves of a black poplar are broad and roughly triangular in shape. They are yellow-green, turning bright yellow in the autumn, and about 5–10cm (2–4in) long. The buds are reddish-brown and pointed. Like the leaves, they are arranged alternately along the twigs – this is typical of poplars and willows.

In common with most other poplars, the leaves of a black poplar have flattened stalks. This flattening causes the leaves to shimmer in the slightest breeze.

Bark and crown The bark of a mature black poplar is a much darker brown than the bark of other poplars – hence the name 'black' – and it is deeply furrowed. There are often large rounded growths near the base of the trunk.

A characteristic of fully-grown black poplars is that they lean away from the prevailing wind, which makes their crowns lop-sided. This is so typical of black poplars that they can often be identified from some distance away by their shape alone.

Easily-worked timber Black poplar has long been commercially important to man because it grows fast and can be felled easily. In the Middle Ages it used to be coppiced in Britain to provide shoots for supporting vines (which were more numerous in the days when the weather was warmer than it is now!) The timber is very pale, with a fine, soft texture. Before efficient cutting tools were developed, the softness of the timber made it a popular, though unsuitable, building material.

Today the timber has limited but specialised uses. Its soft texture makes it easily sliced into veneers by rotating the logs against a sharp blade. The veneers are used for making matches and matchboxes, and are also turned into punnets to hold soft fruits and cress.

Black poplar (*Populus nigra*). Deciduous tree, probably native to Britain. Occurs mainly in central and eastern England where it is often planted in rows to provide shelter or screen a building. Grows to a height of about 30m (100ft).

Opposite: A black poplar in winter. The lop-sided trunk is caused by the prevailing wind and is characteristic of black poplars, allowing the tree to be easily identified from a distance.

Below: The male catkins mature in March before the leaves come out. At first they are grey but turn crimson-red as they release their pollen.

Quick to fall The black poplar's combination of large crown, fast growth and soft timber makes it unstable in a high wind. Large branches can be shed and, in a strong gale, whole trees can be blown over. Not surprisingly, black poplars do not live long and it is often wise to fell a tree before it becomes too old and dangerous.

They are also susceptible to disease, in particular bacterial canker, which attacks the bark and the wood beneath it, causing weeping wounds that slowly kill the larger branches and then the whole tree.

Hybrid poplars The black poplar hybridises easily with certain other poplars, notably the American black poplar and the western balsam poplar. These hybrids grow much more vigorously than the parent trees and quickly reach a considerable height and girth. The best of them have cylindrical stems reaching high up into the tree, with a large quantity of usable timber, and many are resistant to canker. The major poplars being planted commercially nowadays are these canker-resistant hybrids.

The most common hybrid poplar is the black Italian poplar (its cultivar name is 'Serotina'). It is often planted in parks and gardens, as well as for screening and shelter. It resembles the black poplar except that the bark is pale grey and the leaves are reddish-brown when they first emerge in late May. There are only males of this hybrid. The catkins mature in April, becoming bright red by the middle of the month. By the end of April they shed their pollen and have fallen from the tree.

The Lombardy poplar To most people the word 'poplar' refers to the tall, narrow trees often seen planted in rows. These trees are a variety of the black poplar called the Lombardy poplar; so-called because it originated

Right: **Lombardy poplar** (*Populus nigra* Italica). Introduced variety of the black poplar. Height 30m (100ft).

male catkins

Male catkins of the Lombardy poplar (above). Male catkins and mature seeds of the black poplar (below).

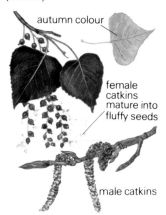

autumn colour

female catkins mature into fluffy seeds

male catkins

Below: The Lombardy poplar has a characteristically narrow columnar habit. Its unusual shape comes from the fact that the branches are all very small and swept strongly upwards. A tree having this shape is said to be fastigiate.

in the Lombardy region of northern Italy.

The Lombardy poplar was introduced to Britain in about 1758. Almost all trees seen here are males–like the black poplar, the female Lombardy poplar produces large numbers of fluffy white seeds that would block up drains and litter pavements.

The unusual shape of the tree is a result of the fact that the branches are all very small and grow upwards, almost parallel to the stem. Its narrow shape and lack of large branches make the Lombardy poplar a useful roadside tree and it is particularly common as such in France and Belgium.

The leaves of a Lombardy poplar are similar to those of the black poplar, but broader and more triangular. The bark is dark grey-brown with shallow ridges. The male flowers are crimson catkins and they ripen in mid-April. The female catkins are green but rarely seen since female trees are hardly ever planted in Britain.

Like the black poplar, the Lombardy poplar is planted to provide shelter and to screen buildings.

Undermining the foundations Poplars are often blamed for undermining the foundations of buildings and even causing cracks to appear in brickwork. However, the problem is much less common than is sometimes supposed and occurs only on heavy clay soils, particularly the sticky London clay. Much of the problem is due to the nature of clay itself. During rainy periods clay takes up a lot of water and swells up. When it dries out it shrinks again. The movement of the clay in wet and dry periods can cause buildings to heave and subside, and if the foundations are too shallow cracks may appear in walls.

Poplar trees tend to exacerbate the problem because they demand a lot of water. During dry periods they make a clay soil even drier than it would be naturally, and so it shrinks more. A particularly bad example occurred during the dry summer of 1976. Since then, building regulations in the areas affected (mainly London and south-east England) have been amended.

WATERSIDE WILLOWS

Our two largest native willows, the white and crack willows, are often deliberately planted by man to help prevent water eroding the soil alongside soggy river banks. They thrive too in damp valleys and fens.

There are more than 130 species of willow in the world, at least 15 of which are native to the British Isles. They include sallows, osiers and many garden species, as well as white, crack and cricket bat willows, three of our most common species.

White willow In its typical form this willow has steeply ascending branches which develop into a narrow crown. The overall grey-whiteness of the tree's foliage gives it its name. The long slender leaves with sharp-pointed tips are light green on top and covered with a thick down underneath, giving the leaves a silvery sheen. The dark grey bark has a close network of deep fissures and ridges and is rich in salicin, a chemical used in the tanning

Below: A white willow growing along the river bank at Wicken Fen in Cambridgeshire. This tree gains its name from the grey-white colour of its foliage. In its typical form, the tree has steeply ascending branches and it reaches a height of about 25m (80ft). The leaves are light green at first, then turn dark green with age; they have white silky hairs on the top and a thick white down underneath.

White willow *(Salix alba)*. Deciduous, native, grows to 25m (80ft). By streams, rivers, marshes, damp woods. Flowers April-May, fruits June.

finely toothed edge to leaf

male catkins (3.5-5.5cm)

(3.5-5.5cm)

White willow

female catkins showing white-plumed seeds

male catkins (2-5cm)

more coarsely toothed edge to leaf

Crack willow

female catkin (showing seeds)

(10cm)

of leather and formerly for making aspirins. The twigs are silky when young and quite tough, in contrast to the crack willow which has fragile twigs.

The dangling cylindrical catkins start to develop in spring at the same time as the leaves. The male catkins produce large quantities of bright yellow pollen which is carried by the wind or insects to the green catkins of the female tree.

Once pollinated, the female catkins develop into small seed capsules which ripen and eventually split, releasing large numbers of white-plumed seeds; these in turn are dispersed by the wind. They lack a protective layer of endosperm (or albumen, a substance like the white of an egg) that seeds of most other tree species possess. They must find a suitable place to germinate within a few days or they die. They need very moist conditions to germinate successfully, which explains why adult willows are so often found in watery surroundings. Once the plants have started to develop, their water requirement is comparatively small.

Crack willow gets its name from the fact that its twigs are extremely fragile and snap easily. The coarsely ridged and fissured bark is dull grey in colour, and the twigs bluish-green. The leaves, which are narrow with coarsely toothed edges and tapered to an asymmetrical point, are glossy green on top and bluish underneath. The catkins are drooping and cylindrical and appear in April at the same time as the leaves and slightly earlier than those of the white willow; male catkins are yellow and female catkins green.

The crack willow spreads partly by the dispersal of its seeds in the same way as the white willow, and also by its twigs. These snap off easily and are carried downstream by the river current; they lodge and readily take root in mud banks or shingle. In common with all willow species, the crack willow grows very easily from cuttings and you often see willow fence posts sprouting new growth, or willow tree stakes outgrowing the trees they were intended to support.

Individual willows are extremely difficult

Above: **Crack willow** *(Salix fragilis)*. Deciduous, native, grows to 25m (80ft). Common in damp places, by streams, rivers. Flowers April, fruits May-June. Below: A peaceful riverside scene – crack willows at Bourne End in Hertfordshire.

to identify and you may have problems deciding which are crack and which white, since they hybridize freely, producing trees which look very similar or intermediate to their parents.

Also, willow leaves can change considerably in shape as they mature from spring to summer, and, confusingly, two willows of the same species may have quite different leaf shapes. (The surest method of identification is to look at the flowers and leaves together.) Each tree produces either male or female flowers (catkins) and you will need to be able to recognise both to distinguish between the species.

Cricket bat willow is the commonest hybrid between the white and crack willow and it probably originated in Suffolk. It can be distinguished from the white willow mainly by its leaves which are grey on the underside. It is extremely fast-growing and its wood is tough, pliable, light and ideal for making cricket bats. One full-sized tree makes at least two dozen bats.

Willow hybrids between other species are cultivated in gardens and towns for their decorative appearance; these include the coral bark, the silver and the golden willow with its attractive variant the weeping willow, probably the most familiar member of the willow family in this country.

Willow timber is a versatile wood because it is light and resilient. Willows were often pollarded to produce straight branches sprouting directly from the top of the trunk, well out of reach of grazing animals. Pollarding was especially important in areas like the East Anglian Fens where little or no other timber was available. It is no longer widely practised. You can, however, often see trees that have been pollarded long ago now growing unchecked. They are common alongside many river banks in the south-east of England.

Host to wildlife Many insects depend on willow leaves for their food. Aphids often cause extensive leaf damage by excreting honeydew, which encourages mould. You also find leaf beetles, weevils, sawflies and gall wasps on willow trees.

Above: Pollarded crack willows near the Thames at Wallingford. The willows in the background have been allowed to grow in their natural dome shape.

DELICATE PUSSY WILLOWS

Sprouting by streams and rivers and in other damp places, pussy willows are most noticeable for their golden catkins, which are nectar pots for hungry bees.

Pussy willows–or sallows–belong to the willow family, many of which can only grow in waterlogged ground; but sallows grow easily in drier woodland and hedgerows. They are small, rounded-shaped trees, quietly beautiful in a way not appreciated by foresters who, when they find them growing in carefully cultivated plantations, treat them as weeds and cut them down.

Male and female Like all willows, sallows bear their male and female flowers on separate trees. The familiar grey pussy willow catkins are the emerging flowers of the male tree that turn golden yellow as their anthers mature. The female trees have less conspicuous catkins, which do not dangle like those of other

Goat willow in the Fens at Easter ablaze with male catkins. In the past bee-keepers used to plant sallows because their abundant nectar was taken by bees to fill honeycombs early in the season.

willows, and become woolly as they mature. Sallows tend to have broader leaves than other willows.

The female catkins are mainly insect-pollinated, with bees their chief visitors. When the fruits ripen and split open they discard hundreds of seeds, each one attached to long silver-white hairs that enable them to be blown around the countryside.

There are three species of native sallow. Finds of fossil pollen show that they have thrived in these islands for more than 100,000 years, surviving at least three glaciations.

Goat willow – or great sallow – is the commonest of the sallows; you find it growing on all types of soil, by ponds and streams, in woods and on wasteland. It gets its name because the young spring foliage was fed to goats. The soft, rounded leaves are grey, with whitish downy undersides. It is cut down so often that it rarely has a chance to grow to its full potential height of about 10m (33ft).

Grey willow – or common sallow – is a smaller, bushy tree which grows on limey, as well as acid, soils. Its leaves are narrower than those of the goat willow, harder to the touch and downy on the upper side. The male catkins tend to be more slender, with paler anthers than those of the goat willow, and the tree bears smaller fruit. The Latin name *cinerea* means ashen or cindery.

The round-eared willow – or wrinkled-leaved sallow – is so called because of the persistent kidney-shaped stipules along the shoots. Its leaves are dull grey-green and wrinkled, with grey downy undersides. It grows as a slight bushy tree or shrub in damp woods, on heaths and moors and beside streams.

Hybrids All three sallows are pussy willows, They interbreed quite easily, so these textbook examples are quite hard to find. Victorian gardeners used to plant a *salictum*, a collection of willows, but they found it difficult to keep accurate scientific records: so many hybrids were produced that their owners tended to mistake them for completely new species.

Sallow uses From Neolithic times about 5000 years ago to the present day sallows have been used to make coracles – small boats covered with skins – and coarse wattles for fending and fish-traps. Sallow stakes used to be used in fencing and sprouted into hedges; if you push a stick of willow into reasonably moist soil, it will almost certainly grow.

From medieval times taxes were receipted with tallies that were usually made of sallow wood. The tally was split with a knife into two irregular halves. The payee, particularly the Government, kept one half on receipt of payment. When the system was abolished in 1826, thousands of old tallies were fed into the boiler furnace of the House of Commons. It overheated and the building was completely destroyed. Today it has a special use as fine drawing charcoal.

Looking at the goat willow

Goat willow/Great sallow *(Salix caprea)* native, deciduous, grows to 10m (33ft). Mainly in damp woods and hedgerows, beside streams and on moors and heaths particularly in E. Anglian fens. Flowers March-April, fruits May.

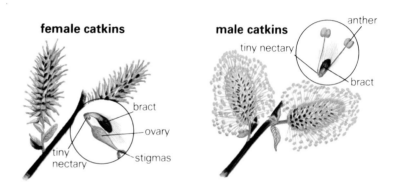

female catkins

bract

ovary

tiny nectary

stigmas

male catkins

anther

tiny nectary

bract

fruiting female catkin

fruit containing many seeds

seed

Identifying the three pussy willows

Goat willow or great sallow *(S. caprea)*

Grey willow or common sallow *(S. cinerea)*

Eared willow or wrinkle-leaved sallow *(S. aurita)*

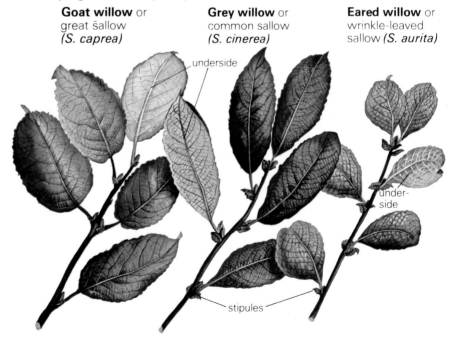

underside

underside

stipules

THE GRACEFUL WEEPING WILLOW

Such an integral part of our lowland landscape is the weeping willow that it is hard to imagine our parks and riversides without it. Yet this tree is not native to the British Isles, being introduced here less than 200 years ago.

Above: The distinctively pendulous shape of a golden weeping willow, in Britain the most widely planted weeping tree of any kind. The tree shown above was photographed in early spring, still in flower.

Two quite separate species of tree are commonly known as the weeping willow. Both were introduced here from abroad – a surprising fact when you consider how naturally suited they are to the riverbanks and parks of lowland Britain. Yet neither species can even be considered as naturalised here, for they cannot propagate themselves on their own; they can spread only with man's help.

Chinese weeping willow The first species of weeping willow to reach Britain is now by far the rarer of the two. This is the Chinese weeping willow (*Salix babylonica*), brought over here during the 18th century. It was named *babylonica* by the Swedish botanist, Linnaeus, who thought it was the tree referred to in Psalm 137:

'By the rivers of Babylon, there we sat down, yea, we wept, when we remembered Zion. We hanged our harps upon the willows in the midst thereof'.

However, it now seems more likely that the tree in question is a species of poplar.

The precise origin of the Chinese weeping willow is still uncertain. It has been in cultivation in western Asia and Europe for centuries, but the plant is not native to these regions and it is thought to have come from western China, spreading westwards via the traditional trade routes.

The Chinese weeping willow made its first

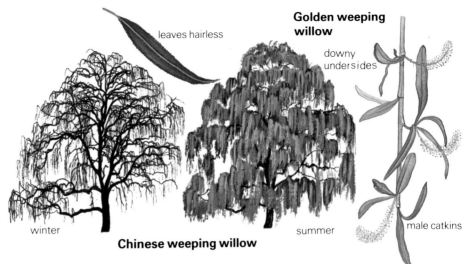

leaves hairless

Golden weeping willow

downy undersides

winter

summer

male catkins

Chinese weeping willow

appearance in Britain when it was planted in Twickenham Park in the 1730s, reputedly by a certain Mr Vernon. A more fanciful story tells that the poet, Alexander Pope, raised the first weeping willows in this country by planting the ropes used to secure a parcel sent from Spain to a friend of his.

New arrival Once introduced, the Chinese weeping willow rapidly became popular and was quite widely planted. However, it has never really thrived in this country because it is not well suited to the British climate. Today, it is rare in Britain, having been superseded by the golden weeping willow.

This species also has an obscure origin, a fact reflected in the number of different names by which it is known, such as *Salix alba* 'Chrysocoma' or 'Vitellina pendula'. Nowadays, many authorities feel that its correct name should be *Salix × sepulchralis* and it is regarded as being a hybrid. Its parents are thought to be *Salix alba* subsp. *vitellina*, a form of the white willow with yellow twigs, and the Chinese weeping willow, from where it gets its conspicuously pendulous habit. It was introduced to Britain from France in about 1800.

Similarities and differences Both the golden and the Chinese weeping willows form graceful trees with long, pendent, recurving branches reaching down almost to the ground. The former grows to a height of 15m (50ft), while the latter can reach a little higher – up to 20m (65ft). The bark of the golden weeping willow is brownish-grey and criss-crossed by shallow grooves and ridges. That of the Chinese weeping willow tends to be more brownish and deeply grooved.

Both trees have the same form of leaf: narrow and lance-shaped with a pointed tip, fine teeth and a short stalk. On the golden weeping willow the leaves can be up to 10cm (4in) long and the lower leaf surface bears very fine hairs, making it appear whitish in contrast to the shiny pale green of the upper surface. The leaves of the Chinese weeping willow tend to be shorter and lack the downy undersurface. The leaves appear very early in the year, usually in February, and remain on the tree until late in the year, sometimes not being shed until December.

Yellow catkins In common with other members of the willow family (which includes poplars and the aspen as well as willows themselves), the flowers of both species of weeping willow are borne in catkins, males and females appearing on separate trees. In both willows the catkins are yellow.

On the Chinese weeping willow the catkins are about 2cm ($\frac{3}{4}$in) long and the female flowers have a nectar-producing gland, called a nectary, to attract pollinating insects – moths and bees being particularly attracted. The pollen grains produced by this and other willows are extremely small and light, light enough, indeed, for the Chinese weeping willow to be pollinated by means of the wind.

By contrast, only male forms of the golden weeping willow are ever seen in this country. This means that the catkins are always male (though occasionally mixed catkins containing a few female flowers occur). Whether pure or mixed, the catkins are slender and about 8cm (3in) long.

Because of its almost complete lack of female flowers, the golden weeping willow has hardly any chance to set seed – though, being a hybrid, its seeds would most likely be sterile in any case. Consequently the tree can be propagated only by cuttings which, fortunately, take root very readily – as do those of many other willows.

Although the Chinese weeping willow has become a rarity in the British Isles, the golden remains one of our most popular decorative trees. Indeed, the attractiveness of many stretches along such rivers as the Thames and the Cam owe a great deal to the planting of this willow. It is difficult to imagine now that this tree has been in Britain for less than 200 years, so much is it a part of our lowland scenery.

Above: Summer and winter outlines of the Chinese weeping willow, with the catkins of the much more common golden variety. The two trees are very similar. The only major differences are that the leaves of the golden weeping willow are downy underneath and its catkins are much longer than on the Chinese species.

Chinese weeping willow (*Salix babylonica*). Deciduous tree, native to western China. Introduced around 1730, but now rare. Height to 20m (65ft).

Golden weeping willow (*Salix × sepulchralis*). Deciduous hybrid, introduced around 1800. Height to 15m (50ft).

Below: The ridged bark of a golden weeping willow.

BALSAM POPLARS

The three species of balsam poplar found in Britain are famed for the strong perfume that wafts from their buds and leaves on moist, warm summer evenings.

Balsam poplars are ornamental trees that were originally introduced into the British Isles from North America during the 17th century. They owe their name to their buds which are coated with an aromatic resin called balsam. In late spring, as the buds split open and the young leaves unfurl, they give off their delightfully fragrant scent.

Three species have been planted in the British Isles: the western balsam poplar, sometimes known as black cottonwood; the balsam poplar; and the balm of Gilead. They have a number of features in common.

Leaves The leaves, which are arranged alternately along the smooth reddish-brown twigs, are more or less oval shaped – rounded at the leaf-stalk and tapering gently to a slender point. The leaf margins are finely toothed.

The upper surfaces of the leaves are dark shining green, but underneath they are conspicuously white or pale green with a fine network of minute veins covering the downy surface. The two-tone colouring of the foliage gives the trees a characteristic quivering appearance as their leaves are shaken by the wind.

The leaves vary greatly in size from $2.5 \times 2cm$ ($1 \times \frac{3}{4}in$) on the side twigs to as much as $25 \times 12cm$ ($10 \times 5in$) on the fast-growing leading shoots. Balsam poplars, especially the balm of Gilead, tend to sprout shoots or suckers from the base of the trunk and here the leaves may be even bigger – $33 \times 35cm$ ($13 \times 10in$) has been recorded.

Unisexual trees Like willows, balsam poplars are either male or female. The catkins that dangle like lambs' tails appear in April, some weeks before the leaves, so that the foliage does not obstruct the free passage of the wind-blown pollen. The male catkins are dull reddish in colour and about 6cm ($2\frac{1}{4}in$) long, while the greenish female catkins are usually twice as long.

The tiny round seeds within the fertilised female catkins ripen by mid-June. Each seed has a plume of fine cotton-like hairs which helps to keep it airborne as it is carried along by the wind. The seeds are rarely released singly, however, because their hairs tend to tangle together, forming a fluffy mass of seeds clinging to the catkin.

Identifying balsam poplars

Balsam poplar (*Populus balsamifera*). Deciduous, introduced; may grow to 25m (82ft). Sometimes planted by lakes, streams. Catkins April, fruits May.

The Balsam poplar is a native tree of N America where it grows in wet wooded areas in river valleys. In Britain it is a rare ornamental tree, planted in damp parts of parks and gardens. The branches spread upwards to form a narrow conical crown. The bark matures to a dark grey colour, its smooth surface cracked by narrow fissures. The buds at the tips of the young shoots are covered with shiny red-brown scales that smell strongly of balsam.

Balm of Gilead (*Populus gileadensis*). Deciduous, introduced; may grow to 25m (82ft). Sometimes planted beside ponds, streams. Catkins Feb-March.

This tree probably arose in N America as a natural hybrid between balsam poplar and eastern cottonwood. Introduced here in 1755, it is now the second most commonly planted species, flourishing in damp areas of parks and gardens. It is in many ways similar to the balsam poplar, but the branches are more spreading, forming a wide open crown. The leaves are broader, almost heart-shaped, and the young shoots and leaf stalks are covered with fine hairs.

Western balsam poplar *(Populus trichocarpa)*. Deciduous, introduced from western N America; may grow to 35m (115ft). Catkins April, fruits May.

The young trees are rather triangular in shape and the trunk is covered with smooth, yellowish-grey bark which tends to peel off. As the tree matures the bark darkens to brownish-grey and is scored with deep vertical fissures. For a deciduous tree its growth is amazingly rapid: one specimen in Kew Gardens reached a height of 16·5m (54ft) in 13 years, a rate of growth that surpasses all other trees that thrive in the British Isles, including fast-growing conifer species.

Opposite page: The western balsam poplar is the species most commonly seen in Britain. The male and female catkins of this species are borne on separate trees. The female (below) is green and when it is fertilised by pollen from the male it develops white fluffy seeds. Male catkins (right) are a dull crimson colour and are often shed in early April before they have released their pollen.

THE HARDY ALDER: A WETLAND TREE

The alder, which lives on wet ground, can be recognised from afar as it tilts slightly over the water. It lines the banks of rivers and streams, invades soggy marshes and fens and colonizes wet woodland. In fact, it grows wherever its roots can bathe in water and absorb rich minerals.

Before man set about clearing forests and draining the land for farming, the alder flourished in the vast valley swamps that covered much of lowland Britain. Nowadays, although it is still a common and widespread tree throughout the British Isles, only small fragments of alder swamp forest (carr) still exist. Some of the best examples of alder carr are found in the fens of East Anglia—around the reed-fringed lakes of the Norfolk Broads for instance. Here alder forms dense, damp woods which encourage a wealth of moisture-loving mosses, ferns and flowering plants.

Appearance In suitably wet ground the alder grows to a height of 20m (66ft) or more, but in drier ground it may not get beyond a

Right: The alder may grow to a maximum height of about 40m (130ft) when conditions are ideal, but it usually only reaches about 20m (66ft). It has a tendency to develop a rather narrow, conical shape. The alders shown here are growing by a mountain stream in Powys, North Wales.

Below: Alder trees growing beside the River Orchy in Glen Orchy, Argyll, Scotland. The presence of alders indicates an unusually rich soil.

stunted, rather bushy stage. The trunk is tall and straight and covered with rough, blackish bark. Where there is plenty of space, the alder's crown is fairly open and rounded, but if crowded by other trees it tends to develop a narrow, rather conical crown. In winter this can give the tree the appearance of a deciduous conifer; this impression is reinforced at close range when you can see the small woody cones which hang on the tree throughout winter.

Look out for the catkins in spring. These were formed the previous summer in preparation for flowering in March and April, about a month before the leaves appear. The alder carries both male and female catkins on

Left: **Alder** (*Alnus glutinosa*) Native, deciduous, grows to 40m (130ft) in wet places by lakes, rivers and streams. May form pure woods in succession to marsh or fen. Lives up to 200 years. Flowers Feb, fruits Oct.

Right: An alder carr brightened by marsh marigolds at Henley Park, Surrey. Alders may be coppiced every 10-15 years.

Above: The alder is our only cone-bearing broad-leaved deciduous tree. The young alder cones appear with the mature foliage in late summer. Alders retain their leaves well into autumn, and they turn a rather dull shade of brown before falling.

Left: Male catkins and old open woody cones that have shed their winged seeds. Some seeds are borne away by wind and others by water.

the same tree. Male catkins, a dull crimson colour in winter, are made up of a column of densely packed stamens. In spring the catkins extend to almost 5cm (2in), almost double their previous length. The stamens separate to reveal masses of yellow pollen. As the flowers dangle in the breeze the pollen is shaken out and scattered by the wind, and some finds its way to the female catkins.

The female catkins are club-shaped at first, about 1cm ($\frac{1}{3}$in) long, and purple-brown in colour. But after pollination they turn green and enlarge to form a rounded cone which protects the ripening seeds.

The purple-tinged leaf buds, arranged alternately along the twigs, split open in May. The new leaves have a sticky surface coating which acts as a death-trap for insect pests which might otherwise chew the delicate young foliage. When the leaves have fully expanded they are dark green on their slightly wrinkled upper surface, and lighter green underneath. They are rather like a tennis racket in shape; the rounded leaves, with prominent veins and toothed margins, taper at the base to join the leaf-stalks which are 5-10cm (2-4in) long.

By autumn the seed-bearing cones have matured to barrel-shaped woody structures whose scales split open to release the seeds. Each seed has a hollow wing on either side which aid wind dispersal. And, more importantly, since many alders overhang rivers, they also act as floats, keeping the seed buoyant as it drifts downstream, until it lodges in a suitable spot and can germinate.

Many seeds that fall in the water are eagerly eaten by wildfowl such as mallard and teal. And the trees themselves attract flocks of siskin and redpoll, winter visitors from Scandinavia, which roam through the branches pecking seeds from the cones.

The only mammal that is particularly associated with alders is the elusive otter which inhabits undisturbed river valleys. The female otter often makes her holt under the roots of an overhanging riverside alder. The cubs are protected from enemies because the entrance to the holt is situated under the water.

Timber Alders produce a useful hard tim-

ber. It was widely used for submerged piles and supports because the wood is extremely durable under water. Indeed, it is said that most of Venice was built on alder-wood piles. Alder was reputed to make valuable charcoal and so many gunpowder mills sprung up in low-lying river plains where a good supply of alders was near at hand. The trees were usually coppiced every 10-15 years.

The timber has also been used for wood-carving and furniture, and for making the soles of clogs that were formerly worn in the north of England, especially around Lanca-shire. In parts of northern Europe it is used to make plywood.

The freshly cut wood is white but it soon becomes stained reddish when exposed to the air. This is due to the release of a dye from the damaged tissues. A tawny red dye can be extracted from the bark, and green and yellow dye from the female catkins and the young shoots respectively.

Alder roots

Where the river bank has been washed away you can see the alder's reddish roots straggling over the bank or streaming in the current. The smallest roots have rather a lumpy appearance. These lumps or nodules are inhabited by bacteria which form a partnership with the alder, beneficial to both (a symbiotic relationship). In return for sugars manu-factured in the leaves and pumped to the roots, the bacteria absorb large amounts of vital nitrogen from the air spaces in the soil and then pass it on to the tree.

Flowers of river and stream

The flowers of aquatic plants form a tapestry of colour floating on the water surface of rivers and streams, while emergent reeds growing along the margins provide a subtle green backcloth. For the riverbank and water-dwelling community of animals the success of this plant community is fundamental to their survival.

One of the major benefits imparted by water plants to rivers is a cleansing service. They oxygenate the water and in so doing help to purify and nullify the effects of noxious substances. The plant life also determines the variety of habitats available for invertebrates and fish. A marginal reed bed, for instance, will shelter an animal community totally different from that of a lily bed. Whatever the habitat, however, the animal variety increases as plant diversity develops.

Emergent and floating-leaved plants also perform one other major service for the animals: they link the underwater world with the air. For many insects this is vital since the juvenile stages are spent under water while the adults live in the air.

Like all forms of life, particular plant species are adapted to specialised conditions. In the centre of a river channel grow submerged and floating species – crowfoots, starworts and water lilies – while at the shallower margins may be found amphibious and emergent species such as amphibious bistort, great yellow cress and water forget-me-not. The interface between water and land might be abrupt, dry and inhospitable, or it might be shallowly sloping, constantly wet and ideal for marshland and fen plants such as marsh orchids, fritillaries and ragged robin. The margins, too, are alive with colour. The reds of hemp agrimony, great willowherb and purple loosetrife contrast with the paler flowers of water plantain, meadowsweet and water cress – all creating subtle variations that make one section of river different from another.

Left: The cuckoo flower (also called lady's smock) with its dainty pale pink blooms, is a typical inhabitant of wet river meadows in the spring.

CHECKLIST
This checklist is a guide to the flowering plants you will find in or near rivers and streams. Although you will not see them all in the same place, you should be able to spot many of them throughout the changing seasons. The species listed in **bold type** *are described in detail.*

Brooklime
Butterbur
Canadian pondweed
Common water plantain
Cuckoo flower
Giant hogweed
Great yellow-cress
Hemlock
Horse radish
Indian balsam
Japanese knotweed
Marsh bedstraw
Marsh bird's-foot trefoil
Marsh marigold
Marsh thistle
Mimulus species
Nodding bur-marigold
Ragged robin
Reed species
Round-leaved speedwell
Rush species
Sedge species
Small teasel
Snake's-head fritillary
Trifid bur-marigold
Water-crowfoot species
Water forget-me-not
Water-lily species
Water speedwell
Yellow flag

Left: Yellow flag and ragged robin growing together in a riverside water meadow. The banks of rivers and streams provide important 'corridors' along which such plants can spread. One condition for their existence is that they can survive occasional or seasonal flooding.

LILIES ON THE WATER

Many water-lilies thrive in Britain, but most are aliens introduced for ornamental purposes; only four species are native to this country.

Water-lilies are a characteristic feature of many slow-moving rivers, canals, lakes and ponds, and a wide variety of forms and colours occur. Although only four species are native to Britain, producing either white or yellow flowers, a very large number of alien species and their horticultural cultivars have been introduced to artificial habitats from where they have escaped, adding red, violet and pastel pink water-lilies to the range we already have in the wild.

White water-lilies Of our four native species only one has white flowers–the white water-lily. Forty other species of water-lily with white flowers are native to other parts of the world; these exotic plants are responsible for

Above: The petals of the white water-lily are large and numerous; the yellow structures in the centre of the flower are its stamens. Notice the clear vein pattern, with the veins radiating from the centre of leaves (see opposite page), a distinguishing feature of this species.

some of the many cultivated forms so popular today in ornamental pools.

Apart from its white flowers this species can be distinguished by its leaves which are opaque with a waxy upper surface and a distinct vein pattern in which the veins radiate out from near the centre of the leaf.

The white water-lily's size is particularly variable and a small form, often referred to as sub-species *occidentalis*, is common on some

Our four native water-lilies

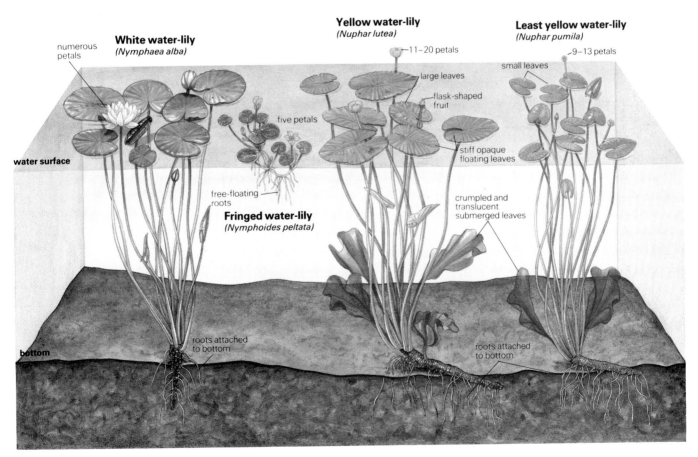

numerous petals

White water-lily
(Nymphaea alba)

Yellow water-lily
(Nuphar lutea)

11–20 petals

large leaves

flask-shaped fruit

Least yellow water-lily
(Nuphar pumila)

9–13 petals

small leaves

five petals

stiff opaque floating leaves

water surface

free-floating roots

Fringed water-lily
(Nymphoides peltata)

crumpled and translucent submerged leaves

bottom

roots attached to bottom

roots attached to bottom

Left: The fringed water-lily is the rarest of our native water-lilies. Unlike other native species it is free-floating and therefore found only in ditches, ponds, canals and slow rivers, where it is unlikely to be washed away. Features that distinguish this species are its yellow flowers, which have only five petals bordered by delicate fringes, and the slightly wavy outline of its leaves.

Below: The yellow water-lily occurs throughout the British Isles except in northern Scotland and the extreme south-west. The flowers have a curious scent resembling that of alcohol, from which (along with its flask-shaped fruits) the plant derives its other popular name—brandy bottle. Like the least yellow water-lily it has two distinct leaf forms: the submerged leaves are translucent and crumpled while those which float on the surface are smaller, opaque and have stiff, waxy upper surfaces.

lakes in Scotland and Ireland, notably Loughs Cregduff and Craiggamore in Galway and Loch Cally, near Dunkeld. Ireland can also claim to have terrestrial water-lilies: in marshes adjacent to some loughs a small form occurs which can grow for long periods out of water because the plants are covered in moisture carried by the prevailing wet westerly winds.

Although normally considered a lowland plant in Britain, the white water-lily has been recorded on lakes above 300m (1000ft). Angle Tarn in the Lake District ((425m/1400ft high), and lakes high up on Isla in Scotland, and in Donegal, Ireland, all support white water-lilies.

The species requires clear, clean water so it rarely colonizes rivers, preferring the more sheltered bays of lakes. Its dependence on good quality water makes it rarer than the yellow water-lilies–two other native water-lilies–but it is more tolerant than they are of brackish water.

Yellow water-lilies The yellow water-lily and least yellow water-lily belong to the same family (Nymphaeaceae) as the white water-lilies but a different genus. Both species have two distinct leaf forms, the submerged leaves being translucent and crumpled and the floating ones smaller and with a stiff waxy upper surface to repel water and help them float. The phenomenon of different leaf forms is known as heterophylly.

Despite these similarities the two species are generally distinct when not in flower because of their considerable differences in size. The yellow water-lily is much the largest, its rhizome often reaching more than 1m (3ft) long and 15cm (6in) across. The leaves are also large and variable, ranging from 10cm (4in) to 50cm (20in) across. In contrast, the rhizome of the least yellow water-lily reaches only about 3cm ($1\frac{1}{4}$in) across and the leaves rarely exceed 15cm (6in). The least yellow water-lily also has a distinct leaf stalk which is more compressed than that of its larger relative, and has an obvious keel.

When the flowers are in bloom the two

species are very different. The yellow water-lily has much larger flowers with 11 to 20 petals, in contrast to the least yellow water-lily which has between 9 and 13. Both flowers are more fragrant than those of the white water-lily and have a pervading aroma of brandy.

The yellow water-lily is the commonest of our four native species and occurs throughout most of the country except in extreme highland areas where it is replaced by the least yellow water-lily. It is the only water-lily to tolerate poor-quality flowing water and so it is a common member of the plant communities found in lowland rivers flowing over clay. It also grows in lakes with rich organic muds devoid of oxygen; the rhizomes, as well as newly formed submerged leaves, can respire anaerobically (in the absence of oxygen). The least water-lily is confined to the pristine clear highland lakes of Scotland, save for two isolated communities in England and Wales.

A most interesting hybrid occurs between the yellow water-lily and the least yellow water-lily; its vegetative and floral characteristics are intermediate between the two parents. The hybrid's main stronghold is in southern and central Scotland where both the parents species occur, but it is also found in Northumberland where only one parent grows, the yellow water-lily.

Fringed water-lily The last of our native water-lilies is a beautiful plant belonging to the genus *Nymphoides* in the bogbean family

Above: The flask-shaped fruits of yellow-flowered water-lilies are unique to these species. They are also important in distinguishing between yellow water-lily and least yellow water-lily. At the top of the fruit lies a flat stigmatic disc which in the former is perfectly etched with 15-20 stigmatic rays that radiate from the centre. Least yellow water-lily, however, has only 8-10 stigmatic rays and a convoluted edge to its stigmatic discs.

(Menyanthaceae)–the fringed water-lily. Unlike other British water-lilies this species is free-floating (not attached to the mud) and is thus confined to ponds, ditches, canals and sluggish rivers. Its leaves differ from those of other native water-lilies in having a slightly wavy outline and only a few veins which arch from the leaf centre and branch repeatedly. Although usually small, the leaves may occasionally be as large as those of the rooted water-lilies. The flowers also bear no resemblance to other water-lilies since they have only five yellow petals bordered by beautifully delicate fringes. Sadly it is a rare plant, occurring only in central and southern Britain.

WATER CROWFOOTS

With their summer blaze of flowers, water-crowfoots are one of the most characteristic plant groups of our rivers, yet telling the species apart can tax even the expert.

Water-crowfoots, or water buttercups as they are sometimes known (they belong to the same genus, *Ranunculus*, as the buttercups), are particularly associated with fast-flowing, shallow rivers. However, different species show clear preferences for different habitats, and they also have distinct geographical distributions.

Yet, despite these variations between the species, water-crowfoots are notoriously difficult to identify. This is mainly because flowing water is a very changeable medium (botanists call it a 'plastic' medium) and the depth of water, the speed at which it flows and its chemical composition can all vary remarkably from one river to another and even within the same river. Such variations, of course, produce a response in the form of the plant, for example in the shape and size of its leaves. A single species of water-crowfoot may, therefore, look totally different depending on whether it is growing in shallow or deep water, in a sheltered pond or exposed to the full force of a river in spate.

Stream species Two species are characteristic of shallow streams rather than rivers, and are known commonly as crowfoots, not water-crowfoots. These are the ivy-leaved crowfoot (*R. hederaceus*) and the round-leaved crowfoot (*R. omiophyllus*). Both are also found in temporary pools and in soft mud at the margins of rivers. Neither develops the finely dissected, submerged leaves of the river species. Instead, the leaves are shaped roughly as indicated by the plants' common names. In flower, the ivy-leaved crowfoot has petals and sepals of the same length, while the sepals of the round-leaved crowfoot are less than half the length of the petals and are also bent backwards down the stalk.

In the wild, both plants are usually annuals and there is no record of their interbreeding, unlike the situation between several species of water-crowfoots. The ivy-leaved species is far more widespread, being found throughout the British Isles. The round-leaved crowfoot is absent from central and eastern England, central and northern Scotland and central, western and the north of Ireland. Yet, despite this, it is the more common species of the two

Above: Stream water-crowfoot in flower on the River Bandon at Inishannon in Ireland.

Left: Water-crowfoots come into flower some time between May and September, the exact time depending on the species. The flowers have the typical structure of a buttercup: a central mass of stigmas surrounded by yellow anthers and a ring of petals, which are white with yellow bases. The species shown here is the river water-crowfoot.

in streams. It is found in small streams at high altitudes, where it prefers water that is acidic and poor in nutrients.

River species The remaining species are all found in rivers and are known commonly as water-crowfoots. There are seven species but you are very unlikely to see more than three species in the same river system because they all have special requirements.

The seven species can be divided into two groups. The first consists of the river, brook, thread-leaved and fan-leaved water-crow-foots. They all develop finely dissected, submerged leaves, but do not bear floating leaves. The remaining three species – the common, pond and stream water-crow-foots can all bear broad floating leaves as well as the submerged dissected kind.

Deep rivers The characteristic water-crowfoot of deep rivers is the fan-leaved water-crowfoot (*R. circinatus*), so called because its delicately dissected leaves are rigid enough to maintain a fan shape under water. This species is typical of Fenland rivers and those flowing over clay soils; it also occurs in canals if the water is alkaline.

The river water-crowfoot (*R. fluitans*) is another species rarely found in small streams, preferring deep water. It roots into the river-bed only at one point upstream of the plant, the remainder whirling downstream in the current. For this reason it requires a very

Below: Brook water-crowfoot (*R. penicillatus* var. *calcareus*) growing in the River Piddle. This species is characteristic of chalk streams with their shallow water and stony, gravelly bottoms. Clear, clean water is essential since water-crowfoots cannot tolerate much pollution. Like the river, thread-leaved and fan-leaved water-crowfoots, the brook water-crowfoot bears finely dissected submerged leaves without the broad floating leaves possessed by some other species.

stable bed to root into.

The river water-crowfoot has a curious distribution. It is common in large English rivers as far west as the Welsh border, and it is occasionally seen in large, stable rivers in Scotland. However, in south-west England and the Lake District it is very rare, and it is not found in any of the Welsh rivers flowing west into the Irish Sea. Yet it does occur as an isolated community in the River Ffraw in Anglesey, and it also occurs in one Irish river, the Five Mile River in Antrim, though it is absent from the rest of Ireland.

The third water-crowfoot of deep rivers, the stream water-crowfoot (*R. penicillatus*), has a complementary distribution to that of the river water-crowfoot, being found in many Irish and Welsh rivers. Although these two species have quite separate distribution patterns in the British Isles, on the Continent they overlap to a great extent. The reason for their British distributions is probably that the river water-crowfoot is a much slower colonizer so, after the last Ice Age, the stream water-crowfoot travelled westwards faster and established itself in Ireland and Wales.

The two species look very similar to each other in their vegetative (ie, non-flowering) state, the stream water-crowfoot being almost as large. In flower, however, the stream species nearly always produces large semi-circular floating leaves, which serve to distinguish it from the river species.

Plants of shallow rivers In chalk rivers, the characteristic member of this group of plants is the brook water-crowfoot (*R. penicillatus* var. *calcareus*), a species seen at its best in the shallow chalk streams of Hampshire, with their crystal-clear water and gravel bottoms. The brook water-crowfoot forms roots at its nodes, which give it a greater ability than the river water-crowfoot to grow on stony, unstable riverbeds. The rooting nodes also allow the plant to colonize sites downstream.

The three species in this group most difficult to tell apart outside the flowering season are the thread-leaved, common and pond water-crowfoots. Fortunately, however, they are not often found growing in the same place. In shallow, nutrient-rich streams feeding into

Right: The British distributions of some of our native water-crowfoots. Most occupy quite distinct geographical areas so that one species is not often seen with another along the same stretch of water. The fan-leaved water-crowfoot (*R. circinatus*), for example, is confined to the Fenlands and rivers flowing over clay soils, while the stream water-crowfoot (*R. penicillatus*) is found only in the far west of the British mainland.

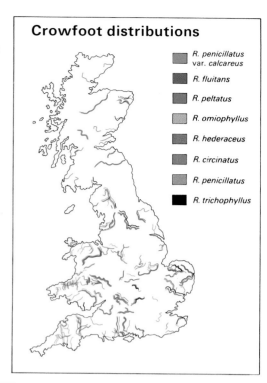

Crowfoot distributions

- R. penicillatus var. calcareus
- R. fluitans
- R. peltatus
- R. omiophyllus
- R. hederaceus
- R. circinatus
- R. penicillatus
- R. trichophyllus

Below: This round-leaved crowfoot is less widespread than the ivy-leaved crowfoot in shallow streams and temporary pools.

Underwater flowering

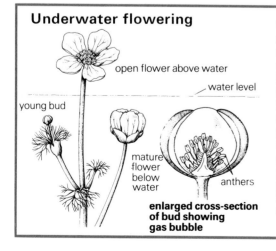

open flower above water

water level

young bud

mature flower below water

anthers

enlarged cross-section of bud showing gas bubble

Water-crowfoots in deep rivers are unlikely to flower if the water level is too high for the plant to break through to the air. A few species, notably the fan-leaved water-crowfoot, overcome this problem by flowering under water, forming a gas bubble within the partly opened bud so that pollination can occur inside a dry mini-environment.

large rivers, the thread-leaved species (*R. trichopyllus*) is the most likely one to be encountered. The pond water-crowfoot (*R. peltatus*) is most typical of the upper reaches of chalk streams, and the common water-crowfoot (*R. aquatilis*) is found in rivers that are poor in nutrients or very alkaline.

The thread-leaved water-crowfoot tends to have the smallest flowers (the petals are less than 6mm long) and the pond species the largest (petals more than 10mm long). But the only sure way to distinguish them is to look at the nectar-producing glands at the base of each petal. These are circular on the common species, pear-shaped on the pond and shaped like a half-moon on the thread-leaved.

RIVER-SIDE MEADOW FLOWERS

In spring and early summer, river-side meadows are ablaze with a profusion of colourful flowers: golden yellow marsh marigolds, rich blue scabious, deep pink ragged robin, white sneezewort and sometimes the rare chequered purple fritillary.

Above: The colourful flowers of ragged robin (*Lychnis flos-cuculi*) resemble those of the red campion, but the petals are deeply cut into four narrow lobes at the margin, creating the ragged look described by the common name.

Some of Britain's richest areas of grassland, in terms of the number of species that grow there, are those found on the moist alluvial soils of river valleys. Like most areas of grassland in this country, these valley meadows are artificial in the sense that they were originally created by man and, if left completely untended, would eventually revert to the woodland that once covered much of our islands.

Changing meadows During recent years there have been many changes in traditional agricultural methods, and one casualty has been the river meadow. It was formerly flooded in winter, then left during the spring to allow grasses and herbaceous plants to flower, but better drainage of the land today has led to the demise of many water-loving species as their habitat is destroyed.

Once the land has been drained it is suitable for ploughing. What was previously summer cattle pasture may make way for grain crops or grasses. If the farmer chooses to cut the existing meadow early for a silage crop, then the developing flowers or fruit are removed. The plants will often try to produce a second crop of flowers, but these too will probably be removed by a second mowing or by grazing. The plant's limited resources are thus rapidly used up and it may not even have sufficient food reserves to survive until the next year. Sometimes meadows – and water meadows in particular –

81

are abandoned as being uneconomical to maintain and the land is soon invaded by more vigorous perennials, scrub and saplings which displace the smaller herbaceous species.

Rare fritillaries Although the flowers that grow in such river-side meadows consist of a varied assortment of species which may be found elsewhere in fens or marshes and other sorts of grassland, some of the flowers found in damp meadows are virtually unknown elsewhere.

One such species is the fritillary or snake's head. The purple flowers of this bulbous species have a chequered pattern that is unique in our flora. More rarely, the flowers are white except for the golden stamens hidden inside the nodding bell of petals. Today, a few sites are maintained especially to encourage the plants to spread.

Pink and red flowers Ragged robin is a more familiar species of damp meadows, as well as of marshes, fens and wood-margins. It has bright pink or rose-red flowers, although white flowers are seen occasionally. The slender base of the petals and the joined calyx lobes form a tube at the bottom of which is the supply of nectar. Because of this structure, access to the nectar is normally restricted to long-tongued insects such as bees and butterflies which pollinate the flowers.

Another species popular with insects–flies as well as bees and butterflies–is the marsh thistle, the commonest of several thistle species growing in damp meadows. The flowers are usually reddish-purple, or occasionally white, and the downy leaves are divided and spiny on the lobes. The young shoots and stems of the marsh thistle, with the spines removed, are used as a salad vegetable in several European countries.

The reddish flowers of the red rattle distinguish it from the similar-looking yellow rattle of drier grasslands. The red rattle is a striking species with purple-tinged stems and foliage setting off the flower spike of purplish-pink flowers, each up to 2.5cm (1in) long. The common name of the species derives from the sound produced by shaking the loose seeds around in the ripe capsules.

Above: The brilliant yellow flowers of the marsh marigold (*Caltha palustris*) are attractive to a great variety of insects, which pollinate the flowers as they crawl about feeding on the abundant pollen and nectar.

Below: Although the stems of the marsh bedstraw can reach up to 1m (39in) in length, they are so weak and slender that they will collapse if the support of surrounding vegetation is removed.

Spring and summer gold The marsh marigold is a common species in wet places, producing rich golden-yellow flowers as much as 5cm (2in) across from March onwards. The flowers of this species are curious in that there is no distinction between petals and sepals and the number of petal-like segments can vary between five and eight, with as many as one hundred stamens clustered inside the petals.

The flower head of the marsh bird's-foot trefoil, a relative of the sweet pea and garden pea, is built up from several pea-like flowers. Each floret is a bright golden-yellow, tinged or veined with red, as are those of its more familiar wild relative, the common bird's-foot trefoil. The species can be distinguished by the more robust, upright stems of the marsh variety. If a stem is snapped in two, the marsh plant has a hollow stem while the common bird's-foot trefoil has a solid stem. It also generally grows in clumps, while the marsh bird's-foot trefoil has isolated stems.

One of the more obvious yellow meadow flowers, because of its height, is the common meadow rue. Commonest in the south and east of England, although found up to northern Scotland, it reaches up to 1m (39in) high and bears dense dark green foliage and masses of fluffy yellow flowers. Close examination of the flowers reveals that the petals are actually white, but so small that the conspicuous cluster of stamens gives the flower

its predominantly yellow colour. The plant spreads by means of underground stems called rhizomes so that large clumps are built up, the thick foliage suppressing the growth of other surrounding species.

White flower heads Sneezewort, an attractive little plant of damp meadows, produces flattened, branched heads of creamy white, daisy-like flowers. These each measure 1.5-2cm (½-¾in) across, and the plant looks like a larger version of yarrow, a related species.

Another white flowered species of riverside meadows, usually found close to free-flowing water, is the marsh bedstraw. It has minute flowers, each with four petals at the end of the stems. Small, backward-pointing stiff hairs on the weak stems and whorls of leaves help the marsh bedstraw to hold on to the surrounding plants.

Water-side blue Sky-blue or pinkish-blue flowers with a yellow 'eye' indicate the presence of water forget-me-nots. These are common in damp grassland as well as by still or running water. Two species are found in river-side meadows. One is *Myosotis scorpiodes* with large flowers measuring up to 1cm (⅜in) across. *Myosotis laxa*, also known as the tufted forget-me-not, has flowers half that size, but is slightly more widespread.

The flower heads of devil's-bit scabious are made up of attractive blue, tubular florets which are visited, and pollinated, by butterflies and bees. The plant's common name

refers to its stout rootstock which ends rather abruptly, as though it had been bitten off. Many folk tales have explained away this characteristic, one of the most popular being that the devil was envious of the considerable range of herbal properties attributed to the species, and so sought to destroy it by biting off its root.

Left: The delicate flowers of the water-forget-me-not appear in marshy areas from May to August.

1. Sneezewort (*Achillea ptarmica*). Flowers July-Aug. Ht to 60cm (24in).

2. Common meadow rue (*Thalictrum flavum*). Flowers June-Aug. Ht to 120cm (47in).

3. Marsh bird's-foot trefoil (*Lotus pedunculatus*). Flowers June-Aug. Ht to 70cm (28in).

4. Devil's-bit scabious (*Succisa pratensis*). Flowers June-Oct. Ht to 100cm (39in).

5. Red rattle (*Pedicularis palustris*). Flowers May-Sept. Ht to 60cm (24in).

6. Marsh thistle (*Cirsium palustre*). Flowers July-Aug. Ht to 150cm (58in).

7. Fritillary (*Fritillaria meleagris*). Flowers Apr-May. Ht to 50cm (20in).

Horse-radish
(Armoracia rusticana)
flowers May-June by
streams, in fields.
Ht 125cm (50in).

Great yellow-cress *(Rorippa
amphibia)* flowers
June-Aug in damp
places.
Ht 120cm (48in).

Right:
Brooklime
*(Veronica
beccabunga)*
flowers May-
Sept in wet
places.
Ht 60cm (24in).

Left:
**Celery-leaved
crowfoot**
*(Ranunculus
sceleratus)*
flowers May-Sept
in damp muddy places.
Ht 60cm (24in).

ADAPTABLE WETLAND PLANTS

Some plants can grow alongside streams, in ditches, damp meadows and soggy woodland, and equally well in fields and on waste ground. Horse-radish and butterbur are good examples of this type of adaptable plant. One is famous for its roots; the other has a fascinating sex life.

Waterside plants live in two environments at once: some parts of the plant are submerged in the water, which is often fast-flowing, and other parts are in the dry air. So it is not surprising that such plants may develop two or more leaf types to suit the different conditions. When a single plant species has many types of differently shaped leaves, this phenomenon is known as heterophylly (from the Greek word for 'different leaves').

The plants illustrated here are inhabitants of pond margins or stream banks, and their type of heterophylly is less spectacular than that of deeper water plants.

Great yellow-cress is an example of a plant which exhibits variety in leaf-shape, size and degree of toothing. This is dependent on the position of the leaves on the stem and the environment in which they grow. The only constant is that leaves at the base of the plant are on short stalks, whereas those higher up have no stalks at all. Horse-radish too, even when growing on dry ground, has comb-like lower leaves and entire upper ones. Divided leaves are probably better able to resist buffeting when submerged.

Horse-radish thrives in the muddy conditions of streamsides, its stout, deep roots spreading easily through the wet soil and throwing up new shoots as they go. It was originally a Middle Eastern plant, introduced to Britain in the 15th century and now quite at home here. It can be found not only by streams but on most types of waste ground, by fields and on roadsides. In fact it is often so prolific that if it is growing in a garden it soon spreads and chokes plants with less sturdy roots.

Although horse-radish is now used exclusively as a condiment, a hot liquid made from the roots used to be drunk to keep out the cold. For this reason it is often found growing near old coaching inns.

Brooklime leaves make rather a bitter salad but are rich in vitamin C. In sunny weather the small blue flowers are normally insect-pollinated, but on dull days they close and self-pollinate when the stamens touch the stigmas.

Butterbur

Towards the end of April, after the flowers of butterbur have withered, the enormous leaves appear and the reason for the plant's name become clear. It probably derives from the practice of using the leaves to wrap butter. But although the leaves are remarkable, sometimes reaching one metre (3ft) in diameter and often smothering large areas of swampy ground, the flowers are even more interesting.

The butterbur belongs to the daisy family (Compositae) and, like the common daisy, its rather complicated-looking flower head is composed of clusters of small flowers. Each daisy flower head contains male and female parts in the individual flowers; but in most British butterburs the female parts are missing and the plants are therefore unable to produce fruit.

In a few counties–Lancashire, Yorkshire, Cheshire and Derbyshire–butterburs with differently structured flowers also occur. The inflorescences on these plants are composed of mostly female flowers which are fertilised by pollen from the male flowers of nearby plants. The pappus hairs around the fruits help in their dispersal.

Why are female butterburs so restricted in their distribution, and why should this species contain two different types of plant?

The answer to the first question may be that male and female plants differ in their requirements. The males may be more adaptable and therefore have managed to colonise a larger area of Britain than their female counterparts.

The answer to the second question seems to be more complex. Many plants have evolved systems of separating the male and female flowering parts to ensure cross-pollination. Some plants such as the hazel have both flowers on the same plant. Others, like holly, produce either pollen or berries, but not both on the same plant. The butterbur may be at an inbetween stage of evolution.

Butterbur *(Petasites hybridus)*. Male and female flowers appear March-May in damp meadows, woodland and beside streams. Female plant only found in Lancs, Yorks, Cheshire, Derbyshire. Ht 150cm (60in).

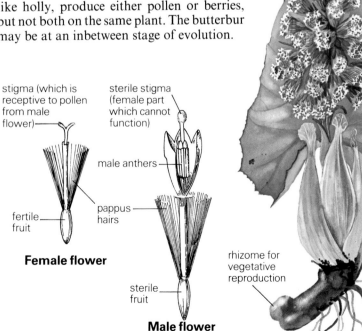

stigma (which is receptive to pollen from male flower)

fertile fruit

Female flower

sterile stigma (female part which cannot function)

male anthers

pappus hairs

sterile fruit

Male flower

rhizome for vegetative reproduction

GARDEN PLANTS ALONG OUR RIVERS

Many of the most colourful plants growing alongside our rivers and streams are not native to Britain but exotic species that have escaped from gardens. Some of these are so successful in the wild that they have ousted our own plants.

Below: One of the less common but most spectacular garden escapes is *Montbretia*, a gladiolus-like plant with deep orange, funnel-shaped flowers. In many rivers of some hillier parts of Britain it occurs only sporadically. Certain other rivers, however, have been invaded so successfully that several native plants –such as sharp flowered rush and floating sweet-grass – were driven out.

For millions of years rivers have provided an ideal pathway along which plants could successfully disperse. Indeed, the natural distributions of many plants today can still be seen to follow water courses. As well as determining to a degree the natural ranges of many native plants, rivers and streams have also allowed exotic plants to spread through the countryside and become established in the wild–a process that has accelerated over the last century as more and more foreign species are brought in to Britain for the horticultural trade. Some of these introductions have been so successful that several of the more beautiful plants, such as *Mimulus,* are almost accepted as our own, while a few have ousted some native species.

Among these 'invaders' some can be said to be truly naturalised while others have to be regarded merely as garden escapes. The distinction between the two, however, is sometimes subtle. A naturalised plant is an introduced species which finds either that the conditions in its new home are favourable to its success or that it can adapt quickly enough to its new environment. It usually reproduces freely and often spreads out to colonize new areas. Naturalised plants found in or along rivers are very rarely localised to a particular area, and most spread extremely effectively. Garden escapes, on the other hand, are usually highly localised, most species being associated with towns, villages or even solitary homesteads. Away from such human influences, garden escapes are rarely encountered along river banks.

Aquatic invaders Compared with the host of introduced species to have colonized river banks, few truly aquatic species have succeeded to the same degree. One spectacular exception, however, is Canadian pondweed.

This species is a native of North America; how it was introduced to this country is still not known. On the mainland of Britain, it first appeared in the wild in the lakes of Duns

86

The spread of giant hogweed in County Durham

Until 1960, only two colonies of giant hogweed were known in County Durham—one along the River Tees, first sighted in 1944, and the other along the River Wear, sighted ten years later. Then in 1960, a colony was discovered growing alongside the River Wear in Durham. Since then, new sightings of giant hogweed have been made in many other parts of the County.

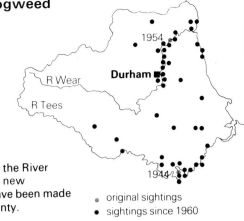

- original sightings
- sightings since 1960

Left: Giant hogweed, with the flat, umbel-shaped head of white flowers so typical of members of the parsley family. This species is a plant to be wary of since it is covered with sharp poisonous hairs that can easily sting anyone who accidentally comes into contact with the plant. If the infected skin is then exposed to sunlight a reaction occurs and a painful blister is formed, which usually takes a very long time to heal.

Castle, Berwickshire, in August 1842. Five years later it was reported to have spread into the streams feeding the River Whiteadder and three years after that the river was choked with it, as was the River Tweed, into which the Whiteadder flows.

By 1852 Canadian pondweed had spread to a great many rivers, canals and reservoirs in Britain, causing serious flooding and navigation problems, especially in the Great Ouse catchment. The problem became so bad in the Fen District that the Government of the day appointed an official to look into the best way of counteracting the menace. Within a few years Canadian pondweed had spread throughout the greater part of Europe.

After its initial domination, Canadian pondweed receded to an almost negligible level before re-invading and stabilising its present population by the beginning of this century. Since then, most other successful invaders have shown the same pattern of initial domination followed by recession and then stabilisation.

A new pondweed In the past few years a ripple of concern, tinged with an uncertain excitement, has travelled through the botanical world with the discovery in southern Britain of a close relative of Canadian pondweed, *Elodea nuttallii*. This species, too, has spread quickly, having invaded all English counties within 20 years. Fortunately, however, its spread has not been as rapid as that of Canadian pondweed and, because it appears to have a more limited range, it may well

prove to be less troublesome. Only a few other aquatic invaders are found in Britain, among them some cultivated water-lilies.

Upland rivers Among the most colourful of our naturalised streamside plants is a group of American invaders belonging to the genus *Mimulus*. These plants are found in most upland rivers with shingle banks. There are four species, each with a characteristic distribution, and a vast range of hybrids.

At the highest altitudes the most common member of this group is a hybrid between the blood-drop emlets (*M. luteus*) and the monkey flower (*M. guttatus*). The flowers of this species are often intermediate between the blotched red and yellow ones of the former

Spreading pondweeds

Female plants of Canadian pondweed (*Elodea canadensis*), shown above with *Elodea nuttallii*, were introduced to Ireland from North America in 1836. Their stems can quickly reach as much as 3m (10ft), producing such dense growth that they can block drains and obstruct waterways. First recorded growing wild on the British mainland in 1842, the species spread rapidly over the next few decades through much of central and southern England. Although this spread has now slowed, the accidentally introduced *Elodea nuttallii* is spreading even more rapidly and within the next decade it may have more or less replaced Canadian pondweed.

Canadian pondweed

- introduced 1842
- 1843–1870 rapid spread
- 1993 limits of spread

Elodea nuttallii

- first noticed 1974
- 1993 present distribution

Above: Four species of *Mimulus* have so far become established in the wild. One of the rarer species is the delicate *M. cupreus* which is common in certain parts of the Lake District and along the banks of some Pennine rivers.

Below: The River Brathay above Lake Windermere is the site of several naturalised plants, including the blue iris (shown here), the Japanese polygonum and the fern *Onoclea sensibilis*.

species and the dotted yellow flowers of the latter. Both parents may also be found in this habitat, the blood-drop emlets being common and widespread while the monkey flower is rarer but also widely distributed.

Another *Mimulus* species, musk (*M. maschatus*), is widely distributed but never common. It prefers rich soils and shaded, wooded river margins. A fourth species, *Mimulus cupreus*, is a delicate copper-coloured plant which is common in some parts of northern England.

Several other introduced plants have successfully become naturalised in wet, rocky places next to moorland rivers–New Zealand willow herb, for instance. But none has done

so well as *Mimulus*. Garden escapes are also rare in upland rivers because of the sparse human population (and hence their gardens) of this habitat.

Invaders of lower reaches As fast-flowing upland rivers flow down to their flood plains their banks gain a much richer mixture of native and naturalised species. An example of a spectacularly successful invader along the lower reaches of many rivers is Indian balsam. This species is now found in most rivers in England, Wales and the middle and south of Scotland, and it is still spreading. Its success is due to its great fertility–a prerequisite for an annual plant. It also tolerates a wide range of soil types, from heavy waterlogged clays to quick-draining shingle islands. Indian balsam is a tall plant, up to 2m (6½ft) in height, with bright pink flowers.

Another continually spreading invader is the giant hogweed. It has a similar distribution to Indian balsam but it is certainly less common. It, too, is a tall plant–even taller than Indian balsam. Only rarely do specimens not attain a height of 2m (6½ft), and occasionally you can see huge plants nearly 4m (13ft) high. Apart from its height, giant hogweed looks like a typical member of the parsley family.

Three other species are commonly naturalised along fast-flowing rivers, though they have not yet reached invasive proportions. The beautiful pink flowers of the North American plant, *Montia sibirica*, are now quite plentiful along shaded stretches of many rivers. The round-leaved speedwell has also spread along many rivers, despite its inability to set viable fruits. The third species is Japanese polygonum or Japanese knotweed.

A more typical example is the fern *Onoclea sensibilis*. This species occurs in abundance along the lower reaches of River Brathay above Lake Windermere.

Garden shrubs such as *Spiraea* are often seen growing alongside fast-flowing rivers near urban areas, while bamboo is a not uncommon sight near very grand houses. Escapes from fruit gardens may also become successfully established. Gooseberry, black currant and red currant bushes all succeed along limited stretches of river banks.

Plants of sluggish rivers No bank-side plants of slow-flowing rivers have managed to emulate the success of some of the garden escapes found along fast-flowing rivers – such as Indian balsam.

One species, sweet flag, is characteristic of a particular type of habitat – the margins of very sluggish rivers flowing over clay. It is, therefore, rarely encountered outside southern England. Within this area it can be common, for example along some stretches of the Great Ouse in Cambridgeshire and the Thames in Berkshire. Sweet flag is native to southern Asia and North America. It was introduced to Europe in 1557 and became naturalised in this country by 1660. In its habit of growth sweet flag is similar to our native flag iris, except that its crushed leaves smell of tangerines and its inflorescence consists of a mass of tiny pale yellowish-green flowers.

The presence of such a variety of naturalised plants from abroad has brought a great deal of extra colour and splendour to our river banks. Yet, the addition of more aquatic and marginal plants should not be encouraged 'willy nilly' in case they become the invaders of tomorrow and displace our own species.

This is another large herbaceous invader which may be increasing in range and number. It produces large leaves which often form a dense cover that seems to be exploited by otters looking for protection.

Lowland garden escapes Besides the naturalised plants mentioned so far, many garden escapes have succeeded in establishing local populations along the lower reaches of fast-flowing streams. The most classical example of an isolated population is that of a species of horsetail, *Equisetum ramossissimum*, which is confined to the banks of the River Witham in Lincolnshire. It has been known to occur there for more than 40 years, yet it has not been found elsewhere in Britain.

Above: A stand of bamboo, one of the more exotic-looking garden escapes you may see along river banks.

Right: Ornamental plants are not the only ones to escape from gardens. Several species of culinary plants, such as horse radish, shown here, have also found their way out of gardens.

Below: Culinary angelica now grows along the damp margins of lowland rivers, often alongside our own native angelica.

WETLAND FLOWERS

From summer to late autumn, streams, ditches and ponds abound with masses of colourful and varied plant life – including some species which are more familiar in gardens and parks.

The luxuriant growth of marsh plants proves that they are suited to their environment, and are able to obtain a good supply of moisture and mineral salts. The marsh environment is rather special, because of the low amount of oxygen in the soil, as there is water rather than air between the soil particles. Certain species – such as the bur-marigolds – have compensated for this, in a variety of ways.

Trifid bur-marigold is an annual with branched, purplish-red stems and large 3-lobed leaves (hence the name 'trifid' meaning cleft into three). The spongy internal structure of the stem contains aerating tissue, which compensates for the lack of oxygen available from the soil, by storing the available oxygen in the air spaces. The brownish-yellow flower heads are about 2cm (¾in) across, and are surrounded by two rows of bracts, the outer ones being green and leaf-like, and the inner ones brownish-yellow. Although a member of the daisy family, the small, dry fruits are not typical, as they are not covered with hairs, but have two or three barbed bristles instead. When animals or people brush by the plant, the fruits catch on to their fur, or their clothes, and are dispersed in this way, falling to the ground some way from the parent plant.

Nodding bur-marigold has thicker stems and undivided leaves with numerous teeth, but is adapted to wetland life in the same way as the trifid bur-marigold. The flower heads have different forms, some have broad yellow rays, and others do not. The name relates to the way in which the flower heads move easily in the slightest breeze.

Rhizomes In the swampy part of a pond, the ground is submerged at all except the driest times of the year. This is the home of tall plants, that have developed long, creeping rhizomes. These rhizomes anchor the plants, and spread under the mud, allowing them to overwinter and produce an abundance of new plants in the spring, ultimately crowding out other species.

Rushes and reeds typically have rhizomes, as does the common water plantain, which often forms large masses in and around ponds. Its thick, tuberous rhizome gives it

a firm hold in the mud, and it is also capable of growing in salt marshes. The leaves grow from the base, but have long stems which hold them above the water level. The whitish, or pinkish-white, flowers have three sepals and three petals, and open only in the afternoon. The unusually shaped one-seeded fruit has a spongy wall, which traps the air, so making the fruit buoyant and helping its dispersal.

Balsams Other wetland plants that have ensured their survival, and indeed increased their numbers by their specially adapted methods of seed dispersal, are members of the Balsam family. Indian balsam, or Himalayan balsam, was introduced to Britain in 1839 from Asia; within 60 years it had become naturalised in many areas, having escaped from the greenhouses where it was originally kept. Its succulent hollow stems and the midribs of the green lance-shaped leaves are reddish. The flowers vary in colour from white and pale pink to a deep rose pink. The plant is often called 'policeman's helmet' because of the shape of the flower, or 'jumping Jack' because of the way in which the ripe seed capsule explodes, due to the tensions set up in the fleshy fruit wall. The 4-12 seeds in a capsule may be thrown as far as 180cm (6ft), a medium-sized plant producing as many as 800 seeds.

Yellow balsam or touch-me-not, is a much rarer and smaller annual. It has succulent stems and swollen nodes (the point at which

Above: The common teasel, despite its prickly appearance, is very attractive to bees.

Right: **Small teasel** (*Dipsacus pilosus*) flowers Aug-Sept in damp woods and ditches. Ht 70cm (28in).

Centre: **Common water plantain** (*Alisma plantago-aquatica*) flowers June-Aug on mud, in or beside water. Ht 60cm (24in).

Above right: **Michaelmas daisy** (*Aster novi-belgii*) flowers Sept-Oct on banks and wasteland. Ht 80cm (31in).

Above centre: **Trifid bur-marigold** (*Bidens tripartita*) flowers July-Sept by ponds, streams and lakes. Ht 35cm (14in).

Above, far right: **Nodding bur-marigold** (*Bidens cernua*) flowers July-Sept by ponds, streams, and in marshes. Ht 35cm (14in).

the leaf appears on the stem). The bright yellow flowers have small brownish or orange spots in the throat.

Hemlock, a member of the carrot family, is a poisonous plant growing in damp areas. All parts of the plant are poisonous, especially the young leaves (which resemble parsley), and the unripe fruits. The ten wavy, ribbed seeds are similar to those of anise or caraway, and the roots are parsnip-like. If you eat hemlock by mistake, respiratory problems or death by paralysis can result.

Usually classed as a biennial, hemlock over-winters as a rosette of the parsley-like leaves. In the summer it bears a flat-topped, creamy white mass of blooms. The smooth, hollow flower stems have dark purple spots on them and the whole plant has an un-pleasant smell.

Garden escapes Some species of wetland plants such as golden rod and Michaelmas daisy, were originally garden plants, or are actually cultivated.

Canadian golden rod was introduced from North America by John Tradescant in 1648. It is an extremely invasive garden escape, and has such a mass of densely packed roots that nothing can survive among them. The lance-shaped toothed leaves are pubescent (covered with fine short hairs) on the underside. The mass of golden yellow flowers provides good bee pasturage, especially in late autumn.

The Michaelmas daisy is another member

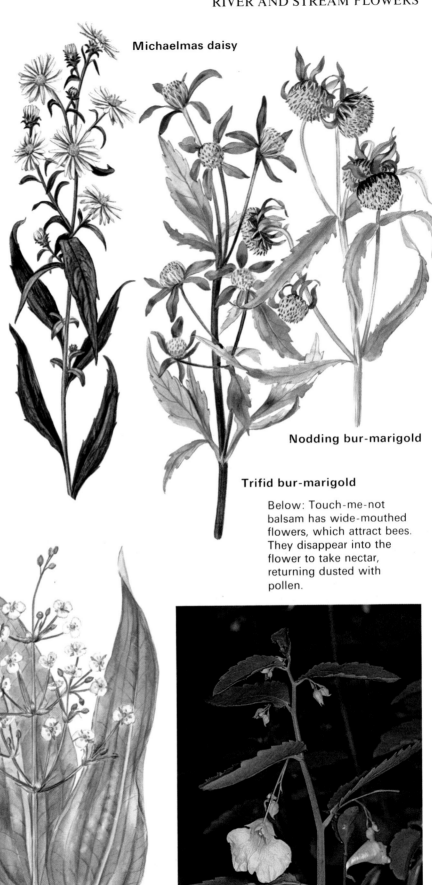

Michaelmas daisy

Nodding bur-marigold

Trifid bur-marigold

Below: Touch-me-not balsam has wide-mouthed flowers, which attract bees. They disappear into the flower to take nectar, returning dusted with pollen.

Small teasel

Common water plantain

of the daisy family, like the bur-marigold and golden rod. It is mainly a garden escape and is found in all colours and sizes, since plant breeders have produced many different varieties from the original violet form – introduced in 1710 from North America.

Fuller's teasel is a cultivated biennial with an upright ribbed stem, armed with short prickles. The opposite toothed leaves are sessile (without a stalk), and a cavity is formed between them and the flower stem in which dew and rainwater collect. Possibly some is absorbed by the plant, but more likely it forms an impassable barrier for ants, which are nectar thieves.

The pale mauve flowers with their prickly bracts are borne in a cone-shaped head. They open a few at a time, from the bottom of the head upwards. The dried, fruiting heads with their stiff, hooked bracts are used in the wool trade to raise the nap of woollen cloth.

The common teasel is the non-cultivated form of fuller's teasel. They differ only in that the seed head spines are straight and longer in the wild form, making them flexible and unsuitable for combing cloth.

The small teasel is a hairy biennial, forming a rosette of long-stalked basal leaves. The small white flowers with their dark violet anthers are borne in a roundish head, and even the fruit is hairy.

Above: Hemlock (*Conium maculatum*) produced the poison drunk by Socrates, the Greek philosopher, sentenced to death in 399BC. The plant may grow up to 210cm (84in) by streams and road sides.

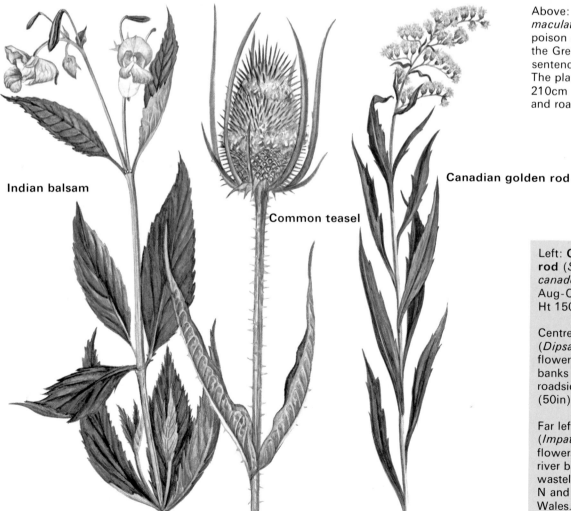

Indian balsam

Common teasel

Canadian golden rod

Left: **Canadian golden rod** (*Solidago canadensis*) flowers Aug-Oct on wasteland. Ht 150cm (58in).

Centre: **Common teasel** (*Dipsacus fullonum*) flowers July-Aug on banks of streams, roadsides. Ht 125cm (50in).

Far left: **Indian balsam** (*Impatiens glandulifera*) flowers July-Oct on river banks and wasteland. Common in N and W England and Wales. Ht 150cm (58in).

RUSHES, SEDGES AND GRASSES

It is quite easy to dismiss sedges, rushes and grasses as apparently rather dull look-alikes. But careful inspection in wet areas such as rivers, streams, ponds, marshes, bogs and fens reveals a number of individual species, each with clearly distinguishable characteristics.

Above: A stand of common reed. The stems of this grass—the largest in Britain—are tough enough to be used for thatching. It flowers between August and October.

On land which is flooded or boggy for the greater part of the year, plants with narrow leaves and inconspicuous flowers make up the bulk of the vegetation: these are the rushes, sedges and grasses. Being wind-pollinated, they lack the attractive colours and scents of insect-pollinated flowers.

Rushes The flowers of rushes are green or brown. The inflorescence, the part of the plant bearing the flowers, is either at the tip of the stem or a few inches below it; on close inspection, each individual flower can be seen to have six papery petals.

Some rushes have flat, grass-like leaves with distinctive straggling hairs along either edge: these are the wood rushes, common

throughout the British Isles. More familiar, though, are the rushes which grow in huge tussocks in wet ground. With some species, such as the soft rush, the stem can be split open to reveal a soft pith, rather like foam rubber. This was once used for the wicks in candles and rush-lights. The tough leaves are still used for chair and basket-making.

Sedges Although sedges and rushes appear superficially similar to one another they are in fact quite easy to distinguish. A rush stem is cylindrical and contains spongy pith. A sedge stem is often three-sided and usually solid, and the leaves grow in three rows up the stem.

A particularly striking wetland sedge is cottongrass which bears a head of fluffy white hairs. All British sedges belong to the Cyperaceae family and most to the genus Carex. Often all the male flowers are at the tip of the stem. Lower down the stem the female flowers become swollen with seeds.

There are about 80 different types of sedge in the British Isles. Two of the most distinctive species are the carnation sedge which is conspicuous for its bluish leaves, and the greater tussock-sedge which grows in huge clumps. Great fen sedge is easy to recognise for its big grey-green leaves armed with vicious teeth along the edges.

Whereas rushes and sedges are often (though by no means always) found in wet places, the true grasses have colonised a wide

1: **Reed sweet-grass**
(*Glyceria maxima*)
flowers July-Aug in deep
water of rivers, ponds,
canals. Ht 2m (6½ft).
2: **Common reed**
(*Phragmites australis*)
flowers Aug-Sept in
shallow water and
swamps. Ht 3m (10ft).
3: **Blunt-flowered rush**
(*Juncus subnodulosus*)
flowers July-Sept in fens,
marshes. Ht 120cm (47in).
4: **Bulbous rush** (*Juncus
bulbosus*) flowers June-
Sept in bogs, moist
heathland, damp woods,
usually on acid soils. Ht
10cm (4in).
5: **Soft rush** (*Juncus
effusus*) flowers June-Aug
in damp woods, bogs, wet
pastures, particularly on
acid soils. Ht 150cm (60in).
6: **Common club-rush**
(*Schoenoplectus lacustris*)
flowers June-July in wet,
silty areas, rare in Wales.
Ht 3m (10ft).
7: **Black bog-rush**
(*Schoenus nigricans*)
flowers May-June in
damp places and by the
sea. Ht 75cm (30in).
8: **Common cottongrass**
(*Eriophorum
angustifolium*) flowers
May-June in bogs and
fens. Ht 60cm (24in).
9: **Common sedge**
(*Carex nigra*) flowers
May-July in damp,
grassy places and beside
ponds and streams on
acid or basic soils. Ht
70cm (27in).
10: **Lesser pond-sedge**
(*Carex acutiformis*)
flowers June-July in
damp woods and beside
streams, usually on heavy
clay soil. Ht 150cm (60in).
11: **Greater tussock-
sedge** (*Carex paniculata*)
flowers May-June in wet
places out of direct
sunlight. Ht 150cm (60in).
12: **Carnation-sedge**
(*Carex panicea*) flowers
May-June in fens and wet
grassy places. Ht 40cm
(16in).
13: **Great fen-sedge**
(*Cladium mariscus*)
flowers July-Aug in reed-
swamps, fens. Ht 3m (10ft).
14: **White beak-sedge**
(*Rhynchospora alba*)
flowers July-Aug in wet,
acid soils. Ht 50cm (20in).

range of habitats throughout the world, from hot deserts to cold polar regions.

Grasses have hollow cylindrical stems; the flowers do not have petals, but are enclosed in two scale-like lobes called glumes. Our tallest native grass, the common reed, is a useful wetland plant: its stems and leaves provide thatching materials and its creeping stems bind river banks together and help to prevent soil erosion. However, if it is not managed properly it can block up whole river systems.

To identify the common reed pull one of the leaves away from the main stem. At the point where the blade meets the sheath clasping the stem you will see a collar of hairs or ligule. In many other grasses this ligule is membranous rather than a crown of hairs.

In most wet places the sweet-grasses put in an appearance too. They range from the diminutive glaucous sweet-grass to the reed sweet-grass which rivals the common reed in stature. The tip of a sweet-grass leaf is shaped like the bows of a boat, in contrast with the flat tips of reed and canary-grass leaves. Sweet-grasses are luscious plants, eagerly grazed by cattle. Because they are nutritious they are encouraged by farmers.

ligule of
fine hairs

sheath

1   2   3   4   5   6   7

Grasses   **Rushes**

Above: Common cottongrass spreads by underground rhizomes. Cottongrass soon disappears if its habitat is drained.

Above: Soft rush growing at the foot of Cwm Idwal in Snowdonia, Wales. The pith was once used for the wicks in candles and rush-lights.

8   9   10   11   12   13   14

Sedges

Above: Greater tussock-sedge seen here with marsh marigolds. Given the right peaty soil, it grows in clumps up to 1m (3ft) wide.

Fishes of river and stream

The size range of British fish is immense. Minute stickleback weigh only a few grams while monster pike and salmon may exceed 25kg (55lb). This variation in size is matched by the ability of different species to respond to variation in their environment; all but the most grossly polluted of our fresh water areas are capable of supporting fish. About half the British species are large enough to be of interest to anglers and this has frequently resulted in their natural range being increased by artificial means. The present-day distribution of the smaller fish reflects more accurately their natural range because they have not been introduced deliberately into new sites for commercial reasons.

The sporting fish which are caught on the fly (salmon, brown trout and grayling) prefer clear, cool, clean rivers with gravelly bottoms. All spawn in oxygen-rich substrates since their eggs would die immediately if they were smothered by fine silts. Other fish, such as barbel, chub, minnow, stone loach and brook lamprey, also spawn in gravels but their requirements are generally less exacting. These species can therefore breed in a greater variety of rivers at lower altitudes.

The majority of coarse fish, on the other hand, tend to spawn among the cover of water plants. Pike, carp, tench, bream, rudd, roach, ruffe, perch and bass are the most adept at this strategy. Weedy canals, with their muddy bottoms and sluggish flows, are thus frequented primarily by coarse fish, typically perch and roach. Canals are intensively fished and they are usually stocked with commercially bred fish. Populations of fish popular to the angler have certainly developed artificially in canals, whereas smaller fish, notably gudgeon, ruffe, minnow and stickleback, have probably established themselves naturally by migrating from streams connected to the canal network.

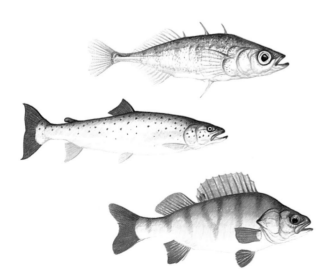

Left: Three examples of river fishes: one of the humblest, the three-spined stickleback; the perch, which is very popular with anglers; and the most magnificent of them all, the salmon.

Left: The stone loach is, perhaps, one of the more difficult fishes to find since it is active only around dusk and dawn. It can on occasion be spotted in shallow riffles where the water runs swiftly over pebbles.

FRESH WATER SMALL FRY

The tiny fishes of streams, rivers and lakes play important roles in the interlocking web of freshwater life, each in its characteristic habitat.

Above: Minnows are fast swimmers and are able to live by a strategy of quick escape. Their survival strategy is also based on staying together in schools so that predators are likely to catch only stragglers.

Opposite page top: The gudgeon, along with the bleak and bullhead, is at the top of the range of sizes in Britain's mini-species at 10-15cm (4-6in) in length. The smallest— the 9-spined stickleback— is only 3-4cm (1¼-1½in).

The word 'fry' can refer to the young fish newly hatched from the egg, or in its first weeks of life. It can also be used as a term for small fishes irrespective of their age— those that might be called the mini-species. So small fry can refer both to the young of the pike, which may grow into a fish of 25kg (55lb) weight, and equally to the fully grown 4cm (1½in) nine-spined stickleback, which is Britain's smallest freshwater fish.

Different communities of small fry occur in different habitats, for it is probably true to say that no two species or age groups have exactly the same requirements in terms of living space, food or spawning places.

Headwater streams Possibly the best known habitat for small fry is the small feeder stream at the headwaters of a river. In lowland areas these lie close to the hills in which the river rises, and may be no more than a metre or two across and 10-30cm (4-12in) deep, with occasional deeper pools at meanders in the stream. In such a situation minnows may be quite common. These live in schools, sometimes numbering hundreds of fishes, in the shallow, sun-warmed water. But they are wary fishes: the sudden appearance of an onlooker on the river bank sends them streaming in a long snake-like school upstream.

Where the water slows at pools, weed beds form, and in the mat of vegetation and plant roots, solitary species such as the stone loach burrow. Here the loach passes the day concealed from light; it is active only in the half-light of dusk and dawn. Loaches can also be found under pebbles in the shallow riffles (riffles are stretches where the water runs swiftly over a pebbly bed).

In eastern England, the locally distributed spined loach may be found in slow streams, but always buried in the river bed or in the

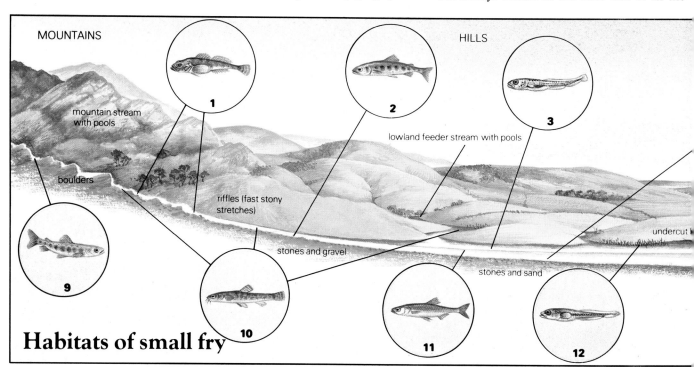

Habitats of small fry

mats of filamentous algae that grow in still areas on the river bed in summer.

Stony riffles–and also dense beds of watercress and other plants–are inhabited by the bullhead, a broad-headed little fish, growing to a length of 10cm (4in), with sharp curved spines on its gill covers. This fish is active mostly at night, and when at rest it lies concealed beneath the larger pebbles. Bullheads even spawn under the pebbles; the eggs are fixed to the roof of their private caves, and the male fish guards them.

Although all these fish are small and of little interest to the angler, it would be a mistake to overlook their importance, or that of the feeder streams. These are, for example, the nursery waters of the dace, which spawn in early March at the tail of the riffles, leaving their eggs among the pebbles to hatch out in about four weeks. Dace fry form small schools in the shallows at first, but soon join together to form larger groups. Without the modest headwater streams, many large rivers would be short of these beautiful, silvery fish.

Trout streams In mountainous areas, streams tend to be more stony and the current fast: not surprisingly, different fishes are found here. Typical of this terrain are little becks not a metre across; in dry weather these contain only a trickle of water running into pools the size of a hand-basin. Like the lowland brooks, these becks contain bullheads and stone loaches, but they are of prime importance as the habitat of trout fry.

Young trout are often segregated by size, the larger ones living in the bigger pools and the smaller ones in tiny pools; but the biggest of them may be no more than 10cm (4in) long, and the individual pool occupied by one of these may be only the size of a kitchen sink.

Below: **1** Miller's thumb– a bottom dwelling species. **2** Salmon parr–lives among gravel on the bottom. **3** Young roach–dwells close to shore in the shallows. **4** Minnows–bottom or mid-water dwellers. **5** The 9-spined stickleback –lives among weed beds. **6** Gudgeon–a bottom dwelling species. **7** Spined loach–a bottom dwelling fish. **8** Bream fry–dwells near the shore region. **9** Brown trout–lives among boulders in pools. **10** Stone loach–a bottom dwelling species. **11** Bleak–dwells near the water surface. **12** Young pike–lives near undercut banks. **13** The 3-spined stickleback –still water with cover. **14** Ruffe–a bottom dwelling species.

As they outgrow their tiny habitats, the trout move downstream to the larger river or lake which the beck feeds. The 'nursery' habitat of their first year is well supplied with insects and other animals, most of which fall in from the banks or are blown in by the wind from neighbouring bushes and trees, but this supply would be totally inadequate for a larger fish. Living in the beck provides a relative immunity from predators, for larger fish-eating fishes and the carnivorous water beetle and dragonfly larvae do not live in such habitats, and even kingfishers and herons can find richer pickings elsewhere.

Small fry in rivers Rivers also provide habitats for species which are more often thought of as still-water fishes. The undercut banks of rivers often have a mat of grass and plant roots, willow stems and brambles hanging in the water. This provides a quiet area out of the current in which the three-spined stickleback thrives. Not only does this environment give protection from the current, but it is also a superb hiding place from fish-eating predators. This is also the

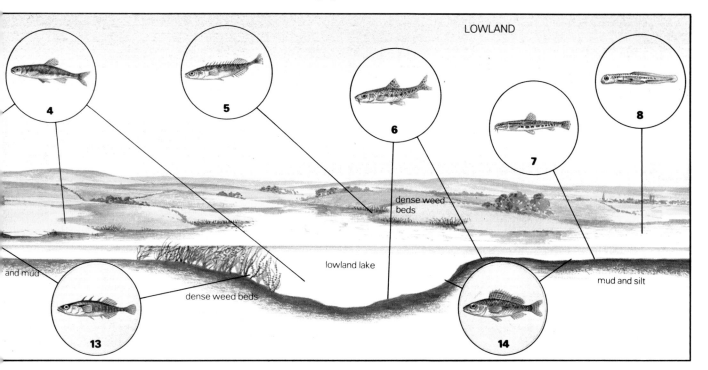

LOWLAND

dense weed beds

lowland lake

and mud

dense weed beds

mud and silt

non-biting midges which are abundant in the rich bottom-mud of slow-flowing rivers.

Similar habitats are adopted by the nine-spined stickleback, which lives in still waters—often those that are very muddy and choked with vegetation and dead leaves. Occasionally this fish is abundant in ditches with only inches of water above the mass of dead leaves, rotting plants and mud—a most unpromising habitat for any small fish. One might expect such habitats frequently to be very poor in dissolved oxygen, so it is no surprise to learn that in experiments this stickleback has been proved to need much less oxygen in the water than its relative, the three-spined stickleback.

The two species of stickleback avoid competing with one another for living space not only by their choice of habitat but also because the three-spined stickleback builds its nest on the bottom of the pond or stream, while the nine-spined stickleback makes a nest among the vegetation a few inches off the bottom.

The gudgeon, one of the larger mini-species in our fresh waters, lives in both lakes and rivers, but in the rivers it is more or less confined to the slow-flowing lowland reaches. As its flattened belly and short barbel at each corner of the mouth suggest, it feeds on bottom-living insect larvae, crustaceans and molluscs, and forms small schools which, when feeding, work along the bottom.

living space of the young pike in its first year, too small at 5cm (2in) or so to be a fish eater.

The minnow and bleak are stronger swimmers than the stickleback, and are therefore better able to live in open water. The minnow lives close to the river bed or in mid-water, and the bleak lives near the surface. Minnows feed on bottom-dwelling crustaceans and insect larvae; bleak feed on planktonic crustaceans, surface-living and wind-blown insects, and spiders which fall into the water. Thus, although both are found close together in the same river, they do not compete for food to any extent. However, living in open water they are vulnerable to predators, and both adopt the same strategy to avoid being eaten: they are fast-swimming and form schools.

The minute fry of such fishes as roach, rudd, dace and minnows are also schooling fish but, lacking the speed of the adults because of their small size, they adopt a strategy of living close to the shore, either near cover or in very shallow water. Although the choice of shallow water makes them conspicuous and more vulnerable to birds like kingfishers, it actually offers protection from the many more underwater predators—pike, perch, trout and chub, and such birds as grebes.

Slow water specialists Small fry in still waters or very slowly flowing streams also include the ruffe. A relative of the perch, this species is at its most abundant in the eastern counties of England. It has the spiny dorsal fin of its cousin, but is a speckled pale brown in colour. It has a series of cavities on the underside of its head which may be sensory in function—evidence for this is the fact that it lives close to the bottom and would clearly benefit from such organs. The ruffe feeds extensively on bloodworms, the larvae of

Above: Trout fry, three months after hatching.

Opposite page: Brown trout live in pools in the quieter parts of upland streams.

Below: A male three-spined stickleback displaying.

GRACEFUL GRAYLING

The grayling – sometimes called 'the lady of the streams' – is a popular sporting fish found in cool, well-oxygenated rivers and mountain streams.

The grayling is a strikingly beautiful relative of the salmon and trout, hence its name 'lady of the streams'. It favours swift running rivers of over 10m (33ft) in width, with deep currents and gravel or rock and gravel beds and is found over much of the northern hemisphere from 40°N to 70°N, with the most southerly sites generally at high altitude. It is also found in lakes, particularly in the north, but also in those with cold water to the south. In northern sites it is tolerant of salt water and can even be found in the sea. Its distribution tends to be rather patchy, but in areas where grayling occur they are often abundant.

British populations The natural distribution of grayling in Britain is very localised – they occur in the River Ouse and its tributaries in Yorkshire, the River Trent, the River Avon and, possibly, the Rivers Severn, Wye, Dee and Ribble. They are not found in Ireland and appear only in the south of Scotland, to a little north of Perth.

Other rivers such as the Test generally contain grayling as a result of introductions made during the last century to provide a greater variety of quarry for anglers. The grayling often shares the water with fishes such as trout, salmon, barbel, dace, chub, roach, eels and minnows. Trout anglers do not welcome grayling in their waters and in many rivers they are systematically removed by netting or other methods in an attempt to reduce their numbers.

As it is sensitive to pollution, the grayling has disappeared from some rivers, such as those in industrial areas. In Britain, it occurs in only a few lakes, such as Llyn Tegid in Wales, and it has been stocked recently in various gravel pits and canals. Unlike its cousins the salmon and trout, the grayling is not an enthusiastic leaper, and its spread has often been limited by waterfalls and weirs.

Feeding habits Grayling feed on insects and their larvae, crustaceans, worms, leeches and molluscs. They also find some food drifting in mid-water or on the water surface. They also graze on the river bed, and on any strands of vegetation, often inadvertently swallowing silt and sand. Some of this may be

Above: The most distinctive characteristic of the grayling (*Thymallus thymallus*) is the large, sail-like dorsal fin. The silvery-grey colouring gives the fish its name. However, bigger fish are darker with red, brown and bluish markings, particularly during the winter and spring. The scales are fairly large and lie in parallel lines, and there is a small rayless fin, known as the adipose fin, between the dorsal and tail fins. This has no known function, and is also characteristic of the grayling's relatives, the salmon, trout and charr.

Opposite: The River Avon in Hampshire is a grayling habitat. The clean, cool water is swift-running and chalky, providing ideal conditions for rapid growth. Although it prefers gravelly river beds with holes and large stones for shelter, the grayling is also found in more open stretches of water.

retained but it is often blown out and any food organisms retaken.

In winter, salmon and trout eggs are often eaten. These may have been washed out of gravel banks by high water, or by salmon and trout spawning where others have already spawned. The large grayling of northern waters often eat other fishes, but this is rare in Britain.

Grayling may form large, well-defined shoals when food is abundant and are often present in considerable numbers. When food is less abundant and fewer fish are present, small loose-knit groups or solitary fish are found.

Reproduction and development Spawning takes place in British waters during March or April in redds on shallow gravel river beds. The male fish wraps his large dorsal fin around the female to delay dispersal of the sperm in the strong water currents. The eggs are quite small, about 3mm in diameter, and they hatch after two or three weeks. When the fry emerge from the gravel, they are 12-18mm ($\frac{1}{2}$-$\frac{3}{4}$in) in length and can be seen in schools in quiet water near the bank. As they grow, the fry move progressively into deeper, faster water. The young fry are easily mistaken for young salmon or trout as they are similar in appearance.

Growth is quite rapid and by winter they attain a length of 10-20cm (4-8in). This is faster growth than most other fishes in the same waters. When the grayling is three years old, measuring 20-40cm (8-16in) and weighing 80-900g (3oz-2lb), it generally spawns for the first time. A few fishes, particularly the males, may mature a year earlier. Particularly fast growth is found in fishes living in chalk streams, such as the Rivers Test and Avon, and in large, rain-fed rivers, such as the Rivers Tay, Annan and Eden. In these large

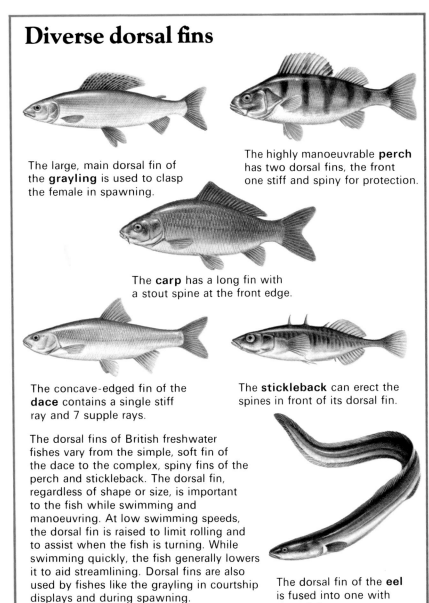

Diverse dorsal fins

The large, main dorsal fin of the **grayling** is used to clasp the female in spawning.

The highly manoeuvrable **perch** has two dorsal fins, the front one stiff and spiny for protection.

The **carp** has a long fin with a stout spine at the front edge.

The concave-edged fin of the **dace** contains a single stiff ray and 7 supple rays.

The **stickleback** can erect the spines in front of its dorsal fin.

The dorsal fins of British freshwater fishes vary from the simple, soft fin of the dace to the complex, spiny fins of the perch and stickleback. The dorsal fin, regardless of shape or size, is important to the fish while swimming and manoeuvring. At low swimming speeds, the dorsal fin is raised to limit rolling and to assist when the fish is turning. While swimming quickly, the fish generally lowers it to aid streamlining. Dorsal fins are also used by fishes like the grayling in courtship displays and during spawning.

The dorsal fin of the **eel** is fused into one with the tail fin.

rivers a shortage of suitable spawning and nursery areas, and the effects of floods and droughts on them, may limit the numbers of grayling, so that the surviving fishes grow rapidly.

In later life, the growth rate slows down and the maximum size of grayling in Britain is about 45cm (18in) in length and 1kg (2$\frac{1}{4}$lb) in weight at five to ten or more years old. In many rivers, the life span is short, with individuals over four years old being rare, and so this maximum size is never reached, although in the past fishes weighing much more than 1kg (2$\frac{1}{4}$lb) were found.

Looking to the future Although it is good for eating, with a distinctive thyme-like smell when fresh, the grayling is not an important fish commercially. However, it is a valuable sporting fish, caught with artificial flies, maggots or worms.

Chronic pollution has destroyed the grayling populations of many rivers in Central Europe and many countries, including France, now take measures to preserve the remaining fishes.

RESOURCEFUL STICKLEBACKS

During the breeding season from March to June, sticklebacks reveal what a male-dominated species they are. Among the three-, nine- and fifteen-spined sticklebacks the male stakes out his territory, builds the nest and looks after the young.

Single-parent family
Among all sticklebacks the males provide shelter and protection for the several broods each one fathers.

The small three-spined stickleback is one of the most common and widely distributed fish in Britain. Few young anglers equipped with jam-jar and dip-net can have failed to catch one since this species lives in virtually all rivers, streams, lakes and ponds, lurking in submerged bankside vegetation within arm's reach. Moreover, as it is easy to catch and can survive in captivity (so long as it is living in an adequately prepared aquarium), it also appeals to naturalists studying animal behaviour. Possibly more has been written about the stickleback than any other fish.

There are eight recognised species (found only in the northern hemisphere), three of which live in British waters. The fifteen-spined stickleback is a sea fish. The nine-spined stickleback is a freshwater inhabitant. The three-spined stickleback seems as content to live in the North Sea as it is to live in a pond on the village green.

Co-existence The three-spined and nine-spined species manage to avoid competing with one another in most aspects of their life cycle, although they share the same habitat. The three-spined stickleback favours open water and is most common in rivers and pools, where clumps of water plants or algae provide cover for the fish to hover around and to retreat to if threatened. It will also take shelter among clumps of underwater plants to avoid the strong flow in a river produced by floods, because it is not capable of strong, sustained swimming.

The nine-spined stickleback, like its relative, is not a strong swimmer. But because it seeks out places where submerged plants are dense and lives among the mat of stems and roots at the edges of the water, it has little need to be. It is rarely caught in open water; and when seen there it makes a quick dash for cover.

So although they may live within a matter of metres of one another, the two species probably rarely come into contact. In the River Roding, for example, a small Essex tributary of the Thames, the three-spined stickleback can be caught in the river in pools where the current is slow. The nine-spined species, however, inhabits a small overgrown ditch which feeds into the river through a patch of swampy bulrushes. Push a net into the ditch, and you will bring up a mass of partly decomposed leaves and black mud with only nine-spined sticklebacks in it.

Such places are prone to dry up in hot weather and the water will contain little dissolved oxygen. So it appears the nine-spined stickleback will survive in water at higher temperatures and at lower levels of dissolved oxygen than its relative–as laboratory experiments have also concluded.

Nest building There are also differences in the stickleback's breeding habits. Each species lays its eggs in a nest built by the male in a territory which he stakes out and guards. The three-spined stickleback is unique, however, in choosing to build its nest–constructed mostly of plant fibres and forming a mass of about 5-6cm (about 2in) diameter–in open ground on the bottom. The nest is usually in a hollow on a patch of sand, or gravel and sand, but on mud if the bottom is muddy.

The nine-spined stickleback constructs its

nest pit

4-5cm

8-9cm

In March the three-spined stickleback claims a territory in shallow open water, defending it against other males until the end of June. He excavates a pit, sucking in sand and spitting it out away from the nest site. Fragments of vegetation are brought to the hollow and 'glued' together with a secretion from the fish's kidneys. This glue is released when the male presses his abdomen against the fragments. Once a nest mound has been gathered he butts it into shape and forces his way through to make a tunnel.
Next, he entices a ripe female to the nest by means of a zig-zag dance, holding his mouth open and spines erect. He quivers against her flanks, inducing her to spawn in the tunnel, then immediately fertilises the eggs. After this he drives her away.

Male spawns with several females until 300-1000 eggs have been laid.

♀

♂

rather smaller nest among the branches and leaves of water plants some 10-15cm (about 4-5in) above the bottom of the lake or stream. Very rarely the male builds a nest close to the bottom. When this happens he excavates a shallow 'doorstep' at the front of the nest to keep the entrance clear of the mud.

The fifteen-spined stickleback builds its nest among algae in a sheltered pool, or below the low-tide mark, but still clear of the bottom. The nest is roughly ball-shaped, about the size of an orange.

Sea-going three-spined sticklebacks show a similar breeding behaviour to those that live in fresh water. They spawn in a shallow pool, usually well clear of strong wave motion, either in an estuary or coastal pool at about high spring-tide level. Their nests are also found on the open bottom, but built of threads of fine *Enteromorpha* algae or detached fronds of seaweed.

The male fish guards the nest alone; eventually he is joined by the young once they are large enough to swim strongly. They form fairly compact schools swimming in the open

water of pools. These pools are flooded by the high tides of autumn and the fish are swept out to sea. Although the fifteen-spined young also form a loose school for the first weeks of their life, they quickly disperse and remain more or less solitary.

Sea-dwelling The sea-going three-spined sticklebacks only occur in the upper regions of the North Sea. They are very abundant— even in mid-ocean. This was shown some years ago when a fine specimen was caught by the crew of an Atlantic weather ship, in a position almost equidistant between Ireland, Iceland, Greenland and Newfoundland. These sticklebacks are not often caught because they live very near the surface, and

Above: Three-spined sticklebacks are most often seen in shallow water keeping station just off the bottom. Here fry and adults alike feed on small aquatic insects and crustaceans.

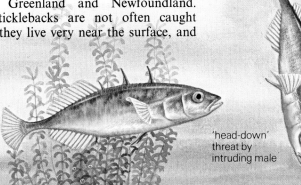

'head-down' threat by intruding male

The male cares for the developing eggs by fanning water around them, so providing oxygen. His vivid colours gradually fade during his parental role.

As eggs develop the male will pull holes in the nest to increase water circulation.

After about eight days the fry (young fish) hatch. The male chases off intruders and prevents his offspring from straying too far. If the fry wander off he gathers them in his mouth and spits them back into the nest pit. When they are 10-14 days old he leaves them to fend for themselves. This pattern is repeated several times during breeding. The nine- and fifteen-spined stickleback story is similar, but differs in detail. The nine-spined fish builds his nest in freshwater weeds and the fifteen-spined constructs his nest in seaweed. Both freshwater fish perform zig-zag courtship dances but the sea fish adopts a tail-biting ritual. In all cases it is the male which is responsible for the care of the young; the female merely lays the eggs.

only special nets will drag the upper few centimetres of the water.

The fifteen-spined stickleback, on the other hand, is entirely a coastal species. It is found from the Bay of Biscay to northern Norway, and in the Baltic Sea. It is equally at home in estuaries as in salty seawater, but not in water deeper than 10m (33ft). It characteristically inhabits the eelgrass (*Zostera*) beds, and sheltered areas where fucoid algae grow. Occasionally it is found in rock pools between the tides.

Spiny protection The three-spined stickleback is by far the most pugnacious of the three species. The fact that it is the only one living and breeding substantially in the open is a reflection of this. Its several long sharp spines, which lock into an upright position on its back and belly, suggest a degree of protection against predators, and certainly the stickleback comes last in the 'popular prey' list.

There are sticklebacks in the Hebrides which, by some genetic fluke, have no spines at all (or, at most, minute ones). Study of their habitat has revealed that there were almost no animals likely to eat sticklebacks. From this it was inferred that they had survived only because there were no predators in the same locality.

Nevertheless large pike and trout eat three-spined sticklebacks, as do otters, kingfishers and a number of sea birds. The other sticklebacks, which adopt a more secretive life style, are not known to fall victim to larger creatures, but this may be because insufficient study has been made of their biology in comparison with the three-spined stickleback.

There are still many questions to be answered about the species, perhaps the most absorbing being what the nearest relatives of the stickleback family can be. For many

Identifying the three species

The names given to the three species provide a clue to their identification, although it is rarely necessary to count the spines to discover which is which. The three-spined stickleback is deeper-bodied than the other two, but still slender. (The pelvic fins also have a long sharp spine each side.) The nine-spined has a rather thin body, but nothing of the eel-like slenderness of the fifteen-spined. As its name suggests it usually has nine spines along the back, although this can vary between eight and ten. They are short, closely spaced, and each has a small triangular fin membrane behind it. The dorsal and anal fins are similar in shape and long-based. The fifteen-spined stickleback is long-bodied with an elongated, pointed snout, and a very slender tail. Both dorsal and anal fins have a dusky patch on the front.

Three-spined stickleback (*Gasterosteus aculeatus*): up to 5cm (2in), but up to 10cm (4in) at sea. First two spines long and quite stout; third is small.

Nine-spined stickleback (*Pungitius pungitius*): up to 7cm (2¾in). Tail is long and narrow, with fan-like fin. Head small but eyes relatively large.

Fifteen-spined stickleback (*Spinachia spinachia*): about 15cm (6in). Tail fin is broad and rounded; dorsal and anal fins are short-based and rounded.

years it was thought that they were related to the scorpion fish, so common in tropical seas, and the gurnards and bullheads which are found round Britain. Later studies pinpointed pipefish and sea horses as the nearest relatives. There is not much similarity between sea horses and sticklebacks, but they have relatives which resemble each other.

Opposite page: Pretty well every river and stream can have a population of sticklebacks—this gravel bedded stream in the New Forest is a likely habitat.

Below: Male three-spined stickleback in breeding dress.

Above: The minnow (*Phoxinus phoxinus*) has a rounded body which in cross-section is rather slender, with a single, short-based fin on its back, a similar but smaller anal fin and a forked tail. Minnows swim in schools for safety, and to spread their search path for food.

MINUTE MINNOWS

The minnow may be small, but it has spread successfully throughout the British Isles –aided by the strategy of swimming in 'schools', and the female's ability to spawn up to 1000 eggs every year.

Why fish form schools

As well as having many pairs of eyes and nostrils – an obvious advantage in detecting food – a school of fish occupies a greater volume which increases its search path. Each member of a school also wastes less effort being on the look-out against predators than a single fish. Schooling fish do not stay in the same position relative to others in the group, and are changing place constantly. A predator selects a fish on the edge of the school, and begins its attack only to find its target fish has been replaced by another – this relay effect causes the predator to pause momentarily and so make a less telling charge. Then on the point of attack, the school may scatter violently, again confusing the predator and helping most of the fish to escape.
All schools have rules: the fish must be about equal size, since swimming ability is related to length, and everyone has to keep up. Sick, injured or parasitised fish quickly drop out – to be taken by predators – because they cannot keep up with the fittest.

The minnow is the smallest member of the carp family. Like others in the group–carp, chub, bream, tench and roach–the minnow has no scales on its head. But it differs from its relatives in having minute scales on the body (most carp fish have large scales), and a lateral line that runs only part of the way along the tail.

Even though the minnow is small, it is an extraordinarily successful fish which is found throughout the British Isles, although it is not widespread in the Scottish highlands nor in the Devon-Cornwall peninsula.

The minnow prefers the headwaters and upper reaches of rivers, where the water has a moderate current, plenty of oxygen and is unpolluted. However, the fish is not fussy in its choice of habitats, and may be caught in large lowland rivers which have clouded water and a slow flow.

Pros and cons of size The minnow's smallness works to its advantage, because it allows the fish to feed on the small crustaceans that are very abundant in most lakes and rivers. An opportunist feeder, the minnow eats whatever animal plankton or insect larvae are most readily available at the time. Larger fish do not feed on these because the total food value of each individual morsel is less than the energy spent to capture it. For the minnow the balance lies the other way.

The minnow's size enables it to forage among stones on the river bed, thrusting into crevices that larger fishes could never reach; it is also an open-water fish, and has much greater opportunities for finding food than such competitors as the loach and the bullhead, which are more or less confined to the bottom. The minnow feeds at all levels in both rivers and lakes–it even captures small flying insects at the surface–but it does not descend much below 6m (20ft) in deeper lakes.

On the other hand, the minnow's size makes it fair game for a number of fish eaters– the kingfisher among birds; perch, young pike, chub and trout among fishes. Carnivorous insect larvae also eat minnow spawn. Once minnows get past this stage they rarely swim singly, forming schools (like other small fish) to keep a look-out for predators.

Spawning success Most minnows become sexually mature at the end of their first year, and spawn early in life. (Larger fish do not usually mature for several years.) The youngest females may produce only 200-300 eggs a year, but they can live up to six years, and mature fish produce nearly 1000 eggs each year.

Spawning takes place from late May to mid-July, usually in shallow, gravelly places such as a riffle (a shallow part of a stream where water flows brokenly) or ford. Lake-living minnows tend to migrate into shallow feeder streams to spawn–though some may spawn on shallow, lake-edge gravel banks. Minnows often form huge schools when spawning and can be approached quite closely from the river bank.

Pimply male The male fish is particularly noticeable at this time of year for its orange-red belly, pectoral and pelvic fins and its dusky black throat.

You may spot a minnow by the conspicuous white pimples (tubercles) on its head. The tubercles or pearl organs, develop only on the males in the breeding season. They are also sometimes known as contact organs. Although many kinds of fish develop breeding tubercles, they are most obvious in the members of the carp family, and especially in the minnow.

These tubercles are composed of keratin – a substance similar to the nails or claws in other animals – which is produced by rapid division of the cells of the outer skin. Tubercles look like small rounded cones. There are many on the male's head, but they are also found on the edges of the pectoral fins where they add a whitish colour.

Courtship colours It is not entirely certain what role the tubercles play in spawning, since the breeding behaviour of the minnow is not all that well known–a paradox considering it is one of our most common fish. As the male minnow acquires bright breeding colours there may be an element of display involved, either before or during spawning–a way to gain the female's attention.

The large white breeding tubercles contrast

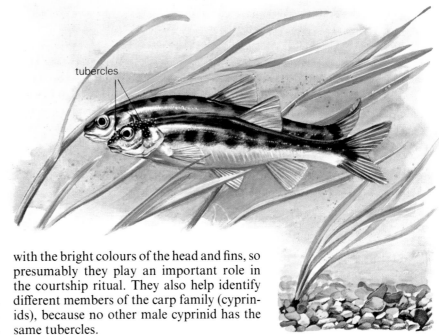

tubercles

with the bright colours of the head and fins, so presumably they play an important role in the courtship ritual. They also help identify different members of the carp family (cyprinids), because no other male cyprinid has the same tubercles.

Spawning ends in summer, and the minnows form small schools comprising fish of about the same length and therefore the same age. They live at or near the surface, dropping close to the bottom at night. In winter minnows live in deeper water. When the river is in spate, the schools disperse and individual fish seek shelter beneath boulders or in bank-side vegetation, and often in shallow water.

Minute terms The word minnow, like tiddler, is often used as a general term for any small fish, and sometimes in a belittling sense for the affairs of Man. The expression 'a triton among the minnows' implies that the triton (part man, part Greek god and part fish), is made to appear greater because of the smallness of his companions.

Many familiar words which suggest smallness also have the same beginning–miniature, minor and minute. Nevertheless, for all its small size, the minnow remains an important link in the freshwater food web.

Above: The tubercles, which appear on the male's head in the breeding season, may assist both sexes to determine their relative position at the moment of spawning. Once they have completed spawning, the male sloughs off his tubercles. Their absence leaves a number of shallow depressions on his head for a few days.

Below: The River Ithon in Powys, Wales is an ideal minnow habitat. The minnow favours clear-flowing, gravel-bedded waters. It also lives in lakes, particularly in hilly regions, though it is not classed as a lake fish.

PERCH: PREDATOR IN THE SHADE

The perch, a popular coarse fish for anglers, is well known for its habit of lurking motionless in shady parts of slow flowing waters. For all this quiet life style, the perch is a fierce predator, attacking its own kind as well as other fish.

Above: The perch (*Perca fluviatilis*) is easy to identify. The broad black stripes on the back, the greenish colouring of the upper sides and the bright orange fins beneath its body are unmistakable. The 14 or so spines in the fin on the back are strong and sharp, as are those in front of the anal and pelvic fins. The perch holds these spines erect when danger threatens. Few predators attack large perch, though herons eat small ones.

The perch is widely distributed in the British Isles, except for northern and western Scotland. It owes much of its present range to introduction by man, especially in Scotland and Ireland. The building of canals has also helped it spread, as have the slow-flowing or still waters that run through much of western England and Wales.

Although it is a strong swimmer, the perch tends to avoid fast currents by lying up in back waters or close to an obstruction in the river. You may often see the fish lying stationary in the water under a bridge or landing stage, or among tree roots under bankside vegetation. It skilfully avoids the water flow by using underwater obstructions to its advantage; hiding in the shade helps the fish to make itself inconspicuous.

Territoriality The perch is not especially active. The young often form schools, but the adults are mostly solitary. Each fish adopts a home range of its own; the size of the perch dictates the area of the range. If a perch is moved from its home range and later released, it will migrate back towards its home stretch of water, travelling as far as 18 or 19 miles in some recorded cases.

Predatory life The perch is a predator from the moment it hatches as a 10-20mm ($\frac{1}{2}$-$\frac{3}{4}$in) long fry. It usually forms schools–groups ranging from 20 to 300–which swim slowly in search of rotifers and the minute, newly-hatched young of copepod crustaceans. These crustaceans, together with bream and crucian carp larvae, are the perch's main food at this early stage. Insect larvae, and larger crust-aceans such as the water slater and freshwater shrimp, later become an important source of food.

If the water contains plenty of food and the early stages of growth have been good, the perch grows to 10cm (4in) by the beginning of its second year. At this size it preys on more young fish, especially members of the carp family and other perch. It takes a heavy toll of the fish spawned a year later than them-selves.

Growth year Perhaps more than any other native freshwater fish the perch shows a tendency towards 'good' years—when many lakes become overcrowded with perch all hatched in the same year. Weather conditions are thought to trigger this off.

If the spring and early summer are particularly warm and sunny, a greater than usual number of planktonic plants (phytoplankton) grow in the water. This means that there is more food for the zooplankton such as the copepod crustaceans, which in turn breed more prolifically than usual. The newly-hatched perch feed on the young zooplankton. More perch fry therefore survive in such a year and a strong 'year class' is established.

In food-rich lakes one year's class will grow large enough to feed on the next year's hatching perch fry. This cannibalism enables a vintage year class to dominate the water for 10-12 years, so making younger perch very scarce. Such unusual feeding habits within a strong year class can, of course, work against the long-term success of the species. Where there is insufficient food, the lake will be inhabited by a large number of rather small perch, all more or less the same age and unable to obtain enough food to grow larger.

Stream of spawn Spawning takes place in warm shallow waters, usually among dense vegetation, submerged tree roots or the twigs or branches of trees that have fallen into the water.

The perch's spawning is affected by temperature. In southern Britain spawning occurs in April, but in cooler, more northern waters it takes place in May. Outside the British Isles, in Sweden for example, spawning may not occur until July. Such variation is due to an in-built timing device which ensures that sufficient food is available for the young fish once they hatch.

A female perch winds a lace-like stream of eggs in a complicated tangle around weeds, roots or twigs. She is accompanied by two or three males who each jockey for position in the attempt to shed their sperm on the eggs.

The eggs hatch in about eight days at 13°C (55°F)—the lower the water temperature the longer they take. After hatching, the perch's rate of growth varies with different water conditions. Five year old fish may reach a length of 27-29cm (10½-11½in) in waters with abundant food, but only 12cm (4½in) in lakes with poor resources. Male perch live for 6-7 years, but females survive longer, from 10-12 years. By the same token, the largest perch, weighing 1.36kg (3lb) and more than 35cm (14in) long, also tend to be females.

Above: It is little wonder that the perch's spawn is known as a stream of spawn —the female may shed up to 200,000 eggs. The number, however, depends on her size; only the largest fish weighing 2kg (4½lb) would lay as many as this.

Below: Fairly slow-flowing rivers are the favoured habitats of perch. These predatory fish establish and defend territories when they are adult.

THE PIKE: HUNTER OR HUNTED?

Few fish have been the subject of taller stories than the pike, our largest native freshwater fish. Exaggerations are also made about the threat of the pike to other water inhabitants.

One of the tallest pike tales comes from Germany and involves the 'Emperor's pike', which was said to have lived for 267 years after the Emperor Frederick II released it into a lake at Lautern in 1230. A fish skeleton of great size, kept in the nearby Mannheim cathedral, was reputedly that of the 'Emperor's pike'. In fact it was later found to have been 'stretched' by the amalgamation of the backbones of two large fish.

Monster pike have been recorded in Britain but often the details are unverified. The 'Kenmure pike' caught around 1774 in Loch Ken in Kirkcudbright weighed 32.62kg (72lb) according to one contemporary author; another says 27.63kg (61lb). When the bones in

its head were compared with another fish from the same loch, they were found to be the same size; the other fish weighed only 17.67kg (39lb). There is no doubt however, that pike in British waters can weigh in excess of 22.65kg (50lb).

The head bones of the 'Lough Conn pike' caught in 1920 and weighing 24.35kg (53lb 12oz) are still preserved in the Natural History Museum in London, although they are no longer on show.

Scarcer and smaller All these large fish were caught many years ago and it is possible that over the last 40 years or so conditions in many of our waters have become less favourable for the pike. More anglers fish for them and, although many anglers do take the trouble to return them safely to the water, some losses must occur. Fishery management practices often call for the removal or decrease in the numbers of pike, as they are predators of other fishes. Pollution has also probably affected the pike as much as any other fish. As a result of all these pressures large pike seem to be much scarcer than they were even half a century ago.

The situation is made worse by the fact that the pike is not a very long-lived fish. It is possible to find out how old a pike is by examining the growth zones of its gill cover (much as one can find the age of other fish from their scales or the age of a tree by counting the rings on its trunk). Study of the

Above: The pike's unmistakable torpedo shape, colour camouflage and large jaws make it a deadly hunter. It can often be seen on sunny days lurking in shallow water among dead vegetation. Typically an inhabitant of lowland rivers and lakes, especially those which contain submerged marginal vegetation, the pike (*Esox lucius*) has been introduced in many areas since it is popular as an angler's fish.

pike in Lake Windermere has shown that it is quite exceptional for these fish to live much longer than 15 years and 18 years was the maximum age recorded in a very large sample.

Where it is possible to be certain of the sex of the fish it has also been found that all the largest pike have been females. Out of a total of 7000 fish taken from Lake Windermere for research into pike, the largest male weighed 5.76kg (12lb 8oz)–the largest female was 16.92kg (35lb).

The size that pike attain and their rate of growth vary very much with the availability of food. Some of the best pike waters are the large lakes where water high in dissolved minerals and rich in nutrients provides a good food supply. Thus growth is remarkably fast; Loch Lomond in Scotland also contains very large pike which have grown relatively fast. The underlying reason is abundant food. The Irish loughs contain large numbers of trout, and other fish such as rudd and bream, while Loch Lomond has powan and other species which provide plentiful nourishment.

Voracious feeder The pike is a predator all its life and because of its size there are few fish in our fresh water which it will not feed on. When first hatched, however, it takes mainly small invertebrates, later graduating to insect larvae and small crustaceans like water fleas. At the end of its first year of life it has grown sufficiently to enable it to take young fish, although it will still eat substantial numbers of aquatic insect larvae. Most of its food consists of fish which are active swimmers, especially shoaling fish such as roach, minnow, dace, perch and trout, but it also eats numbers of sticklebacks if available.

Large pike extend their diet beyond fish and are known to eat ducklings and occasionally fully-grown mallard. Other swimming water birds have been found in their stomachs, as have water voles and brown rats. There have also been reports of attacks on dogs; an alsatian swimming in the River Lea near London, was seized by the tail. Pike do not, however, attack man although many an angler has suffered bites when unhooking a captured specimen; large, live pike have to

be handled with respect to prevent accidents.

Stealthy hunter The pike relies on stealth to capture its prey. It lies up in vegetation where its dappled colouring keeps it well camouflaged, and waits for fish to pass close by. Then, propelled by the huge paddle-like tail end (all its propelling fins are at the end of its body), it makes a lightning charge at its prey. Mostly it hunts by sight; it has a groove, rather like a gunsight, on the snout in front of each eye which helps its forward vision. As a result most of its feeding is done during daylight hours. It has a moderately good sense of smell and can detect an angler's dead bait at night, although finding food by scent is not its prime method of feeding.

Pike prejudice Because the pike is a predator, there is at times prejudice against it on the grounds that it eats too many fish and birds. While there is certainly no place for a pike in a trout farmer's ponds, in most natural waters the pike has a positive role. Simply because pike are fish-eating predators they are always scarcer than their prey and it is estimated that even a lake the size of Windermere can contain no more than 4600 adult fish. In fact we rely on predators to keep in check the number of other species; otherwise the water would become overrun with small, stunted fish. For these reasons there is no justification for killing pike or indeed for controlling their numbers except in unnatural situations.

Above: The pike's fierce looking mouth has a set of needle-sharp teeth in the upper jaw which point backwards to prevent prey from escaping. The pike is adapted to make a powerful seizing lunge and will tackle fish almost its own size, as well as frogs, voles and water birds.

The early stages

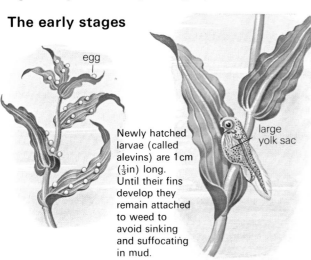

egg

Newly hatched larvae (called alevins) are 1cm ($\frac{1}{3}$in) long. Until their fins develop they remain attached to weed to avoid sinking and suffocating in mud.

large yolk sac

Over 3-4 weeks between late March and early May a female pike may shed up to half a million sticky eggs. Each egg is 3mm ($\frac{1}{8}$in) across.

very small yolk sac

For the next 10 days the larvae remain attached to leaves by an organ on the front of their head, while fins and mouth develop further. 10-day old alevins are 13mm ($\frac{1}{2}$in) long. The yolk sac is very reduced but the mouth is now fully formed.

At 3cm (1$\frac{1}{4}$in) long the young pike resembles the adult: it swims freely and feeds on live food.

BRONZE AND SILVER BREAM

No other British fish forms such huge shoals as our two bream – the common or bronze bream and the rarer silver bream. Thousands of fishes can be found together, all churning up mud as they scour the river-bed for small bottom-dwelling creatures.

Above: At spawning time in spring the males of both the common and the silver bream develop white growths, called tubercles, on their bodies. These occur mainly on their heads but they often spread to the rest of their bodies. The tubercles are rough to the touch and their presence may be connected with the males' practice of bumping and nudging the females prior to spawning.
The tubercles on this male common bream can be seen clearly.

As its name implies, the common bream occurs much more frequently in Britain than the silver bream. It is also known as the bronze bream, yet its body colour is really a dark olive, though there are variations. Younger common bream are often a pale silvery colour, whereas old specimens sometimes become a true metallic bronze; this colour may also vary with the environment.

The other species occurring in the British Isles, the silver (also known as the white bream), is a smaller fish than the bronze. It rarely grows longer than 36cm (14in), compared with an average length for the common bream of 40-50cm (16-20in). Unlike the common bream the silver does not have any colour variations, always being silvery with pinkish pectoral and pelvic fins.

The best way to tell the two species apart is by counting the number of scales on the lateral line and the number of rays in the anal fin: in the bronze bream there are 25-30 rays and 51-60 scales, and in the silver bream these figures are 21-23 and 44-48, respectively.

The fishes hybridise freely when both inhabit the same water, the hybrids being intermediate in colour, scale pattern, number of rays in the fins and so on. Similarly, hybrids occur between both bream species and rudd and roach. In all cases the numbers of fin rays and scales on the hybrids are intermediate between those of the parents.

Ancient arrivals The common and the silver bream are native to the streams of south-east England, from the Thames north to the Trent. They arrived in Britain just after the last Ice Age. At that time, England was joined to the Continent and rivers such as the Thames and the Trent were tributaries of the Rhine. Bream (and many other species) spread down the Rhine and into these English rivers. When England was separated from the Continent with the formation of the English Channel, the bream here were isolated.

From a focus in south-eastern England, the common bream has spread to many other parts of England, and has been successfully introduced to Wales, Scotland and Ireland. The silver bream however, is confined mostly to East Anglia.

The typical habitat for both species is a weedy slow-moving lowland river such as the Nene, the Welland, the East Anglian rivers and the Somerset drains. There are big populations of bream in man-made lakes and reservoirs, old water-supply lakes at the head of canal systems and the Norfolk Broads.

Shoals of thousands No other British freshwater fish forms shoals as readily as the bream. It lives all its life in shoals, some of which can number thousands of individuals. As the members of a shoal get bigger, so their numbers decrease, and bream over 5kg (10lb) in weight may move around in groups of as few as ten.

Shoaling allows the bream to patrol certain areas of a river or lake as a 'home range'. They feed mainly on bottom-dwelling invertebrates such as snails that live within their territory. They locate their food by rooting about in the mud, a technique that releases clouds of particles into the water. In clean rivers like the Thames, you can often see large muddy patches at the surface, indicating a feeding shoal of bream. The fishes often betray their presence by breaking the surface with their dorsal fins and tails.

Feeding usually takes place early in the morning, though it sometimes carries on into the day and even the night as well.

Shallow spawning sites Shoal behaviour is most prominent at spawning time, when vast groups of bream gather prior to breeding.

common bream

silver bream

Above: The common bream is indigenous to southern and eastern England, though it has been introduced to Wales, Scotland and Ireland. The silver bream, being less tolerant of fast water, is mostly confined to East Anglia.

Below: The bream is the typical fish of weedy slow-flowing lowland rivers and canals, such as this disused canal, which is noted for its shoals of spawning bream. Man-made lakes and reservoirs and especially gravel pits often contain huge shoals of this fish. Another good place to find bream is country lakes of the type built by Capability Brown and other 18th century landscape designers.

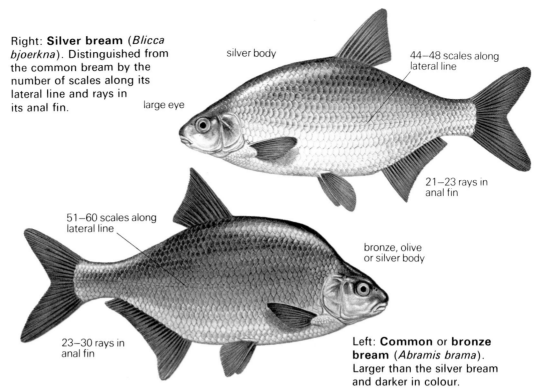

Right: **Silver bream** (*Blicca bjoerkna*). Distinguished from the common bream by the number of scales along its lateral line and rays in its anal fin.

silver body

large eye

44–48 scales along lateral line

21–23 rays in anal fin

51–60 scales along lateral line

bronze, olive or silver body

23–30 rays in anal fin

Left: **Common** or **bronze bream** (*Abramis brama*). Larger than the silver bream and darker in colour.

Spawning takes place in late May when the temperature of the water has reached 14-16°C (57-61°F). The favourite sites are weed-beds in sunny shallows.

Before spawning, the males develop hard growths called tubercles on their heads and bodies, and they take up a territory, typically a piece of weed-bed about 2m (6ft) long. They can be there for anything up to a week before the females arrive.

Spawning normally occurs at dawn or dusk. The females are driven into the weeds by a group of males, and chased around and in and out of the weeds. This physical buffeting and the friction of the weeds, together with the males' tubercles, combine to draw out the females' eggs, which are then fertilised by milt from the males. Being sticky, many eggs adhere to the weeds, but some fall to the bottom of the river or lake, where they often die from a lack of oxygen.

The female bronze bream is very fecund and can produce 90,000 eggs per kilogram (40,000 per pound) of her body weight. The eggs hatch in five to six days at 16°C (61°F), the larvae at first being only 4-5mm long. They stay on the weeds (or wherever the eggs settled) for a few days while they absorb their yolks, but they soon have to find food in large quantities. The young larvae feed on diatoms and zooplankton, and grow fast.

Growth rates The rate at which bream grow depends to a great degree on the environment, particularly on the number and size of other fishes present and on the productivity of the water. At the end of the first year, an average size for a bronze bream is 5-6cm (2-2½in).

The ultimate size of a bronze bream and its long-term growth rate depend on the type of water and the levels of predation and egg production. In lakes of 10 hectares (25 acres) or more bream can grow to be very large and very old. Fishes aged more than 20 years have certainly been recorded and they may survive for much longer than that; the biggest bream caught in British waters weighed over 6.5kg (14lb). In small waters, however, bream tend not to grow as large, and a specimen weighing 1-2kg (2-4½lb) would be considered fairly big.

The bream is one of our most popular freshwater angling fishes, and it can be caught in great numbers if an angler contacts a large shoal in a feeding mood. Much harder is the catching of a very large bream. An angler may fish for a lifetime and never see a 5kg (11lb) bream, and those who do will probably have fished for hundreds of hours to catch it.

MYSTERIOUS TRAVELS OF THE EEL

The remarkable story of the eel starts and finishes deep in the Atlantic Ocean, but how it navigates the thousands of miles to and from Europe remains a mystery.

Above: The common eel (*Anguilla anguilla*) undergoes many transformations in its life cycle. Here, in its yellow eel form, it develops tiny scales so deeply embedded in its skin that you cannot see them.

Below: This grid-iron eel trap has caught silver eels as they try to return to their breeding grounds to spawn. This 'run' of eels only happens on pitch black nights, often after a storm when the water is dirty.

In March and April, adult eels spawn in the deep water of the Atlantic Ocean including the Sargasso Sea north-east of the Caribbean, and then die. Meanwhile the tiny larvae that hatch from these eggs start their long journey of over 3000 miles across the Atlantic to Europe. Here the young eels move into rivers, streams and lakes to develop until, some years later, they are ready to leave their freshwater habitat and return to the sea to breed.

Atlantic crossing Young eel larvae, which are called leptocephali and look rather like transparent leaves, are about 5mm ($\frac{1}{4}$in) in length – quite unlike the long snake-like adults. Indeed, they were once thought to be a separate species of fish until, at the end of

the last century, scientists kept some leptocephali alive in a tank and discovered that they turned into young eels. It was not until the 1920s that the larvae's journey across the Atlantic was plotted in any detail. A Danish scientist charted their growth over the migration route and concluded that the Sargasso Sea, where the smallest specimens were found, must be the eels' spawning ground. However no one has yet proved this.

It is thought that eels lay their eggs in very deep water and in early summer the newly hatched larvae rise to the surface, where they feed on plankton. During their first year they grow rapidly as they are carried across the Atlantic by the Gulf Stream and other currents moving towards the European coast; this migration lasts about three years. Any larvae drifting too far north die.

Elvers in estuaries When the larvae reach coastal waters they stop feeding and change into elvers – miniature versions of adult eels, also commonly called glass eels because they are transparent. As they move through the brackish water of estuaries and into the fresh water of rivers and streams, they soon darken in colour. In the British Isles this movement, which occurs in almost every river, takes place in winter and early spring; a well-known location is the River Severn. Some elvers stay behind in the muddy water of estuaries, feeding on shrimps, worms, crabs and small fish. (There is no truth in the theory that these are males and that only the females move upstream.) The journey up the estuaries is hazardous, with sea birds and fishermen taking their toll.

Travelling upriver Once in the rivers the elvers' progress upstream may be helped by man-made eel passes, which allow them to move over waterfalls, locks, dams and other obstacles in their path. The water flow is checked by a tube of wire netting filled with straw, heather or twigs, which the eels can use as a kind of ladder up and over the obstruction.

By the second winter in fresh water, eels have trebled in length, feeding on small snails and insect larvae. Eels are nocturnal animals, hunting for food at night and hiding

in mud, vegetation or under stones during the day. Those that live in the colder northern areas spend winter hidden in mud or under stones.

Yellow eels Eels continue to grow in the years they spend in their freshwater habitat. During this time they are called yellow eels, because they have a yellow belly; the back is brown. A yellow eel has a soft body, a broad snout and small eyes. These details are worth noting since the eel undergoes yet another change later in its life, and the difference in size and colour between the yellow eel and the final stage in its life cycle used to cause a lot of confusion. Many believed that the two forms were in fact different kinds of eel, until it was discovered they were merely two stages of growth.

Silver eels Exactly when yellow eels become silver eels, with a silvery belly, pointed snout and a hard body, varies enormously and occurs when the eels are anything from four to more than ten years old. The eyes grow bigger, but the reason for this change is not clear. Certainly it is not to assist hunting for food, since they stop feeding at this stage. Males develop into silver eels about two years before the females.

Long haul home Autumn is the time when silver eels begin the migration downstream on their way to the breeding grounds in the Sargasso Sea. Recent research indicates eels are not sexually mature until after they reach

Right: Elvers measure 6·5cm (2½in), shorter in fact than the fully grown larvae. Their journey up the estuaries is not easy. They escape the seaward drag of the tide by burrowing into the sand, but they are vulnerable when swimming, especially to the fisherman's eel net (above). The elvers are trapped as they move against the current and are then taken to eel ponds until large enough to be sold.

Eel migration

adult eel over 6 years (40cm)

← adult migration to spawning area

→ migration of larvae in Gulf Stream

■ newly hatched larvae (7mm) in spawning area

■ larvae 1-2½ years (75mm)

■ larvae under 1 year (50mm)

■ elver 3 years (65mm)

the open sea, although very few eels have been discovered at this stage. The fact that years ago eels' eggs or newly hatched fry were never found caused much speculation about their reproduction, including the belief that eels were sexless and had some mysterious method of breeding.

In research it has been shown that migrating eels cannot be deflected from their seaward route–and that captive eels become restless when river eels start their journey. On their way to the sea they travel along rivers, streams, ditches and other waterways– even overland on dark wet nights, when their thick skin and narrow gill slits prevent them drying out.

Although the migrating silver eels do not feed, they have plenty of stored fat–up to a quarter of their body weight–to sustain them on their journey; captive silver eels can live for several years without food. In the sea, the mature adults reach their destination in the spring, anytime from 6 to 18 months after leaving the fresh water.

Little is known about how the eels navigate across thousands of miles of ocean. Even today, few adult eels have been caught in the sea–and those only in coastal waters; but they may be guided by the increase in temperature and saltiness of the water as they head towards the Sargasso Sea. Once there, the eels lay their eggs, fertilise them and then die. No eel ever returns.

CLINGING LAMPREYS OF RIVER AND SEA

Lampreys – sole survivors of an ancient group of animals – start their life as larvae, buried in the mud of rivers. As adults, they have powerful suckered mouths which they use to attach themselves to stones or unsuspecting victims.

Above: As adults, brook lampreys do not feed and have weakly developed teeth in their sucker discs. They use their discs mainly to attach themselves to stones to resist water flow. Other species of lamprey have obvious, sharply pointed teeth which enable them to take firm hold of their victims. All of the species breathe through their rows of open gill slits while they are attached to surfaces by their sucker discs.

Despite their eel-like appearance and aquatic lifestyle, lampreys are only distantly related to the more familiar fishes of seas and rivers. Lampreys form a group with primitive ancestors known as the jawless fishes, which differentiates them from true fishes which have jaws.

Ancient ancestors There are three species of lamprey found in our waters – the lamprey or sea lamprey, which is the biggest species; the lampern or river lamprey; and the brook or Planer's lamprey. Together with the similar-looking marine hagfishes, these lampreys are the sole survivors of a group of animals known as Ostracoderms, whose ancestry goes back many millions of years to the Palaeozoic Era.

Lampreys have retained the characteristics of this ancient group. Instead of jaws, they have an oval disc at the head end equipped with teeth, the arrangement of

which varies according to the species. Their gills open by a row of seven holes or pores on each side behind the eyes; and they have fleshy fins on the back, which are more or less continuous with the fin around the tail tip. None of them has pectoral, pelvic or anal fins, and there are no rays or spines in those fins that are present. They do not have bones, the skeleton being made up of gristle, and their backbone consisting of a rudimentary spinal cord, known as a notochord, with no individual vertebrae. Their elongated body is quite scaleless and always very slimy.

Distribution The brook lamprey is widely distributed throughout Britain, although it is probably more abundant in lowland areas. The numbers of lampern and sea lamprey have declined in rivers where impassable weirs have been built, or where pollution has extinguished the populations. However, the lampern is still common in the River Severn and many of its tributaries. The sea lamprey is, however, an uncommon animal except in some Scottish rivers.

Early life in the mud Young lampreys are small and eel-like, reaching only 20cm (8in) in length. Their fins appear as weak folds of skin on their backs, and their minute eyes are largely skin-covered. The mouth disc is small and narrow and there is no sign of the conspicuous teeth of the adults. Identification of the young of the different species – often known as prides – is difficult as they are all similar in appearance.

The prides live buried in river mud, where the current is slack – usually in backwaters or eddies. Their food consists of diatoms and other minute forms of life, and possibly bacteria which they suck from the surface of the mud. Their mouth discs contain an intricate filtering mechanism which enables them to separate this food from the inorganic mud particles. This larval stage of their life lasts for between five and eight years, after which they emerge from the mud and undergo a form of metamorphosis in which they become adult lampreys.

Predatory lifestyle The adult lifestyle is also unusual; two extreme forms are found in the three British species.

Both the sea lamprey and the lampern move downstream into the river mouth after metamorphosis and both become parasites on fishes. This is made possible by their sucker disc, which grips tightly to the skin of the intended victim, while the sharp teeth on the 'tongue' in the centre of the disc bore a hole through skin and scales into the flesh. The structure of their primitive gill slits enables the lampreys to breathe while they are attached to their victim by their sucker discs. Special glands in the mouth produce an anti-coagulant saliva so that the blood of the prey flows freely and the lamprey feeds on the body fluids until the fish is weakened. Most fish, made feeble by the attack, recover; but the wound is susceptible to bacterial and fungal infections.

The sea lamprey ranges far out to sea during this feeding stage in its life history. It is often found attached to basking sharks,

Above: A rare view of the larva of the brook lamprey, which normally spends this stage of its life buried in the mud. The larva, known as a pride, appears blind as the eyes are almost covered by skin. The gill openings are separate and distinct, and the sucker disc has yet to develop sufficiently so that it can attach itself to stones. Metamorphosis to the adult form takes place in the autumn.

and scars from its sucker have been seen on dolphins. The lampern, however, probably stays close to the estuary during this feeding stage, attacking many kinds of larger fishes. Instead of going out to sea, it may migrate down river into streams, where it feeds on freshwater fishes.

Non-feeding adult The brook lamprey moves only a little way down river from the area in which it spent its larval life and then returns to a suitable area upstream. This species is not parasitic and, in fact, does not feed at all once metamorphosis has taken place. This accounts for its weakly developed teeth in the sucker disc. It uses this disc only when attaching itself to stones to avoid being carried downstream by the current, or when moving small stones to prepare a spawning bed.

Spawning grounds All three species spawn in rivers in the spring, the two migratory species (the sea lamprey and lampern) sometimes travelling long distances upstream to find suitable spawning grounds.

The lampern and sea lamprey choose a spawning site with a pebbly bottom and coarse sand, usually partly in the shade. They attach themselves by their suckers to medium-sized stones and pick them up, moving them a little way from the nest site. The nest is further excavated by the lampern attaching itself to a larger stone and vigorously flapping its body up and down, thus dislodging the sand which is carried downstream. The small yellowish eggs are laid in the nest and, to ensure fertilisation, the male attaches himself to the female just behind her head, winding his body around hers as the eggs are laid.

Brook lampreys also seek flowing water and partially shaded conditions for spawning. This takes place in April—or later, depending on the water temperature—in small streams, sometimes only 60cm (2ft) across. Brook lampreys are far from shy while spawning and can be seen in many streams at this time of the year. However, it is often the only time they—and the other lampreys—can be seen, for the adults of all three species die after spawning and their progeny spend the next few years hidden in the river mud.

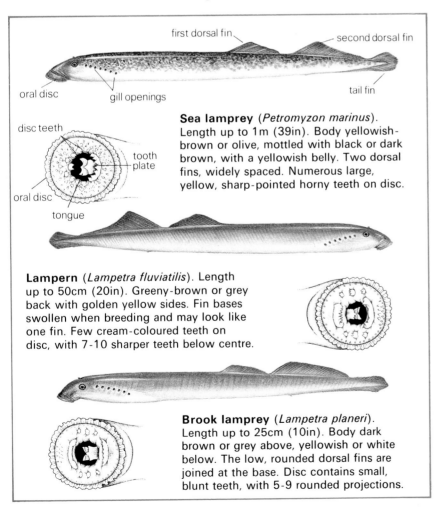

first dorsal fin

second dorsal fin

oral disc

gill openings

tail fin

disc teeth

tooth plate

oral disc

tongue

Sea lamprey (*Petromyzon marinus*). Length up to 1m (39in). Body yellowish-brown or olive, mottled with black or dark brown, with a yellowish belly. Two dorsal fins, widely spaced. Numerous large, yellow, sharp-pointed horny teeth on disc.

Lampern (*Lampetra fluviatilis*). Length up to 50cm (20in). Greeny-brown or grey back with golden yellow sides. Fin bases swollen when breeding and may look like one fin. Few cream-coloured teeth on disc, with 7-10 sharper teeth below centre.

Brook lamprey (*Lampetra planeri*). Length up to 25cm (10in). Body dark brown or grey above, yellowish or white below. The low, rounded dorsal fins are joined at the base. Disc contains small, blunt teeth, with 5-9 rounded projections.

Cold-blood creatures of river and stream

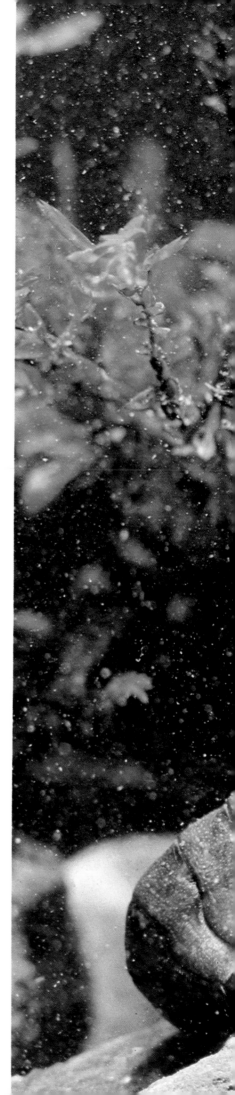

Rivers and canals support a vast array of invertebrates, ranging from the large crayfish to microscopic single-celled zooplankton such as the lowly amoeba. Between these extremes are tiny mites, snails, mussels, leeches, worms and – most interesting of all – thousands of insects of all shapes and sizes.

Dragonflies and damselflies are probably the best loved and observed insects of rivers. In summer the territorial nature of the dragonflies can be seen as they patrol their own patch of water like whirring miniature helicopters; intruders are fought off between bouts of sunbathing at favoured vantage points. The beauty of the terrestrial adult contrasts markedly with the drab and ugly aquatic nymphs, yet both stages are voraciously carnivorous. Damselflies share much in common with the dragonflies but they are much smaller and more delicate, their flights being silent, gentle flits from reed to reed.

An aquatic larva and a free-flying terrestrial adult is a very common lifestyle for many river insects. Apart from the young of dragonflies and damselflies, juvenile mayflies, stoneflies and some bugs are also called 'nymphs'. This term is reserved for larvae which metamorphose into adults without going through a pupal stage. Thus, these insects have larvae which, when mature, crawl out of the water to shed their skins and become instant airborne adults. The mayflies, which belong to this group, are probably best known to fishermen because of their importance as a source of fish food. Mayflies are also unique in the insect world in having two – not one – adult stages. Despite this, mayflies live for only a day.

The other insect group to arouse fascinated study is that of the caddisflies, which rank as the most resourceful of aquatic house builders. Some of the larvae build protective mobile homes of tiny stones, while others crawl around inside tubes of plant stems and leaf debris, disguised as sticks.

Left: The emperor dragonfly is one of the largest and most striking insects to be found flying over the freshwater habitats of Britain. It flies steadily on a regular hunting 'beat' up and down a river or canal bank.

Left: The white-footed crayfish is the only species native to British streams. It favours waters flowing over chalk and limestone soils since it needs the lime to build up its protective outer shell.

121

TERRITORIAL FRESHWATER CRAYFISHES

The white-footed crayfish, found in quite large populations in many lakes and rivers of the British Isles, is a spirited defender of its territorial rights.

Crayfishes resemble diminutive lobsters, and indeed probably derive from the same ancestors. Numerous species and sub-species have evolved since the forerunners of the present day crayfish entered fresh water, and now the richest crayfish populations are found in northern and central America. A limited number of species are found in Asia and Europe, but only one in Britain.

This is the white-footed crayfish (*Austropotamobius pallipes*), which takes its name from the whitish undersides of its walking legs. The white-footed crayfish is our largest freshwater crustacean, growing to an average size of 10cm (4in), although it can reach 12cm (4¾in) in length. It occurs in large numbers in suitable habitats from shallow streams to deep reservoirs. Found throughout the British Isles, it prefers water associated with soft rocks such as chalk and limestone. Waters of the hard and volcanic outcrops of Scotland, some regions of Wales, and Cornwall and the Lake District are too acidic for it to survive. Lime-rich water is important since the crayfish uses the lime to build up its shell.

Feeding habits Crayfishes are omnivores, feeding chiefly on dead animals and some water plants although they are not averse to catching slow-moving fishes or small, sluggish animals including snails and insect larvae.

Above: During September and October, the fall in water temperature helps to provide the stimulus for crayfishes to mate. The male takes up an aggressive posture with his formidable claws outstretched to grapple with any opponent, and also to turn the female on to her back. It seems likely that receptive females liberate a chemical into the water that triggers this behaviour.

Below: The River Darent in Kent, where crayfishes can be found.

Those crayfishes living in shallow streams and rivers forage for food at specific times – at dusk and again at dawn. It may be that the low light intensity at these times offers some protection from predators. During these nocturnal searches for food, crayfishes display territorial behaviour, establishing territories among the stones and weed. A resting crayfish will emerge from its hide to 'challenge' others that approach too close to its territory. Crayfishes held in captivity display similar behaviour, which becomes more apparent when they have to compete for limited shelter, and it seems that they establish constantly changing hierarchies among themselves.

Hazards and diseases The white-footed crayfish moults about six or seven times during its first year, while it is still growing. Until the second moult, it is carried around by its mother, but after this it is very vulnerable at moulting time, and many crayfishes are lost to predators. The average life-span under normal circumstances is six to seven years, and maturity is not reached until the third or fourth year, after which the adults probably breed each year.

The greatest number of mortalities in the crayfish population occur within two to three months of the young hatching. Studies made of stream populations show that some time elapses before the young acquire the nocturnal feeding habit. This is a disadvantage, as their relatively soft bodies and small size makes them ideal food for day-feeding trout and other fishes. The main predators of adult crayfishes are herons, otters and fishes.

Outbreaks of disease occur from time to time among crayfish populations. The white-footed crayfish is sometimes affected by porcelain disease, which is caused by a minute single-celled parasite that invades and breaks down the muscles.

There is now serious concern for the health of the native white-footed crayfish because of recent outbreaks of crayfish plague – a highly contagious fungus that can dramatically reduce or even eliminate whole populations. It is thought to have been carried into this country by the signal crayfish. At present the crayfish populations north to the Midlands are being attacked by the disease, but farther north it has yet to become a problem. Rivers such as the Hampshire Avon are affected for most of their length, yet despite major outbreaks occurring near crayfish farms, there is no clear evidence to link the farms with the disease. The problem for the native crayfish is that it has no capacity to become immune to the fungus, so drastic action will have to be taken if the species is to be saved from ultimate extinction in the near future.

Cultivation In past centuries, the white-footed crayfish was considered to be a delicacy. However, because of its small size when compared with its European relatives, it has never been seriously cultivated, al-

Mating in crayfishes

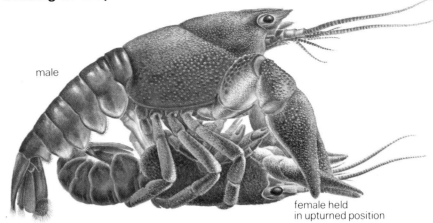

male

female held in upturned position

though in Europe crayfish cultivation has been practised for centuries. Recent interest in the farming of crayfish in this country has led to the introduction of at least one north American species, the Californian crayfish (*Pacifastacus leniusculus*) into a number of private waters. Live red-footed crayfishes (*Astacus astacus*) have also been imported for sale to restaurants. If these species escape into our natural waters in sufficient numbers, they could possibly establish themselves and compete with our native species. These imported species are hardier than the white-footed crayfish, and therefore have much greater resistance to the outbreaks of crayfish plague.

Above: The male turns the female on to her back, then fastens a white, adhesive ribbon of sperm packets on to her underside. These fertilise the eggs she produces later. The females have to be relatively passive during mating since those that are aggressive or reluctant to mate are often injured by the males. Some days after mating, the female is ready to spawn. She turns on her back and lays her eggs into a 'pond' of jelly held in her abdomen.

Right: The eggs are held by the fine hairs on the limbs of the female's abdomen and she carries them for about nine months, until the following June. During this time she hides among submerged tree roots and beneath stones in deeper water. When the crayfishes hatch they measure slightly less than 1cm ($\frac{2}{5}$in) in length. They moult as they grow and after the first moult, attach themselves to their mother's limbs, using their small claws (below).

CREEPING FRESHWATER MUSSELS

From tiny pea mussels to pearl producers, there are 28 species of freshwater mussel to be found in canals, rivers, lakes, and even ditches and drains all over the British Isles.

Freshwater mussels are molluscs, with many features in common with their close relatives, the marine oysters and mussels. They are all bivalves – their bodies being enclosed in two flap-like shells known as valves. These valves are formed from secretions of the mussel's outer skin (the mantle), and extend as two sheets on either side of the body.

Slow-motion swan mussels The five larger species of freshwater mussel are the most frequently found. Of these, the swan mussel is the largest – growing up to 22cm (8¾in) – and most common, occurring in canals, lakes, ponds and reservoirs all over the British Isles, apart from the extreme south-west of England, and the north of Scotland. Like all

mussels, it obtains food and oxygen from the water which enters the mantle cavity through a special aperture, the inhalant siphon, while the valves are partly open. The clean oxygen-bearing water flows through the two lattice-like large gills on either side of the body. The surfaces of the gills are covered with thousands of microscopic beating threads called cilia, that cause the water to flow in through the siphon and over the gills, bringing in oxygen and taking away carbon dioxide. The gills are ideally placed to sift small particles of food, such as microscopic plants and animals, out of the water. These are trapped in slime on the gills and swept towards the mouth by special ciliated pathways. The swan mussel therefore does not need to move to catch its food. All the sustenance it needs is pumped in with the respiratory water.

Movement is possible in all the mussels by means of a blade-like foot which the mussel forces out through a gap in the open valves to lever itself slowly over the river bed. The mussel pushes forward the foot to make a furrow in the surface of the riverbed, and expands the tip to take a firm hold on the furrow wall, then shortens the foot hauling itself forward. This has to be repeated many times before the mussel moves any appreciable distance, and it has been calculated that it would take a freshwater mussel one year to move a mile. However, in times of

Above: The inhalant siphon, the frilled edge of which is visible here, is always exposed in the swan mussel *(Anodonta cygnea)*. A hinge at the top of the shell holds the two valves together, but it has to work against an elastic ligament which acts like a compression spring, forcing the valves apart. The mussel uses two sets of powerful adductor muscles – one at the front and one at the rear – to hold the valves closed. They work against the ligament, allowing the animal to protect itself by closing the valves completely. When the mussel wants to open the valves, it simply relaxes the adductor muscles.

drought, movement is necessary so that the swan mussel can leave the drying shallows to find a new location in deeper water.

Mussel reproduction In the breeding season (June to August) the male swan mussel releases sperm into the water near the female and some of this is drawn into her mantle cavity as she feeds, fertilising the eggs. The female stores the fertilised eggs in the outer folds of her gills, using them as brood pouches. Here the fertilised eggs develop into larvae, known as glochidia, which are miniature bivalves with a tooth on the edge of each valve. The larvae remain in the brood pouches for up to nine months, then they are expelled, clinging to weeds or in small clusters on the mud. When a fish swims by the larvae attach themselves to its fins or skin. They are carried around by the fish in this way, living parasitically until they are old enough to lead an independent life on the river bed.

The duck mussel is similar in so many respects to the swan mussel that its exact status as a separate species has been doubted. It is distinguished from the swan mussel by its slightly smaller size, and a shell which is more oval and swollen, particularly towards the front. The duck mussel is more widely distributed in Scotland than the swan mussel, and prefers moving water and sandy shallows, where it can burrow.

Artists' accessory Artists once used the shells of painter's mussels as containers for their paints – hence their name. They are longer and narrower than the swan and duck mussels, and they also differ in having teeth on the hinge of their shells. These are used to keep the two valves correctly aligned. The painter's mussel, which also reproduces by means of glochidia larvae, is found in slow-flowing rivers and canals over most of England and mid-Wales.

Pearl producer The pearl mussel sometimes produces valuable pearls inside its shell. Although in Roman times pearl fisheries were quite common, this mussel is now only locally fished commercially. It produces the pearl from nacre (mother of pearl) which lines the shell and is secreted to surround a foreign body, such as a sand grain, which has intruded into the shell. As it prefers fast-flowing soft water it is found mainly deep in the swift rivers of south-west and northern England, Wales and Scotland, where it burrows into the sand that accumulates in the lee of large boulders.

Marine relative The zebra mussel, its shell ornamented with a pattern of alternate, wavy, zig-zag bands of brown and yellow, is more closely related to the marine species of mussels. Found in slow-flowing rivers, canals, reservoirs and docks, it attaches itself to solid objects such as tree roots, pilings and the underside of barges, using sticky threads called byssus threads, that are a characteristic of marine mussels. These are

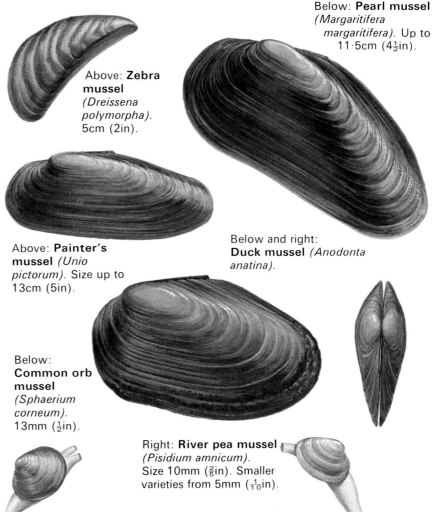

Above: **Zebra mussel** *(Dreissena polymorpha)*. 5cm (2in).

Below: **Pearl mussel** *(Margaritifera margaritifera)*. Up to 11·5cm (4½in).

Above: **Painter's mussel** *(Unio pictorum)*. Size up to 13cm (5in).

Below and right: **Duck mussel** *(Anodonta anatina)*.

Below: **Common orb mussel** *(Sphaerium corneum)*. 13mm (½in).

Right: **River pea mussel** *(Pisidium amnicum)*. Size 10mm (⅖in). Smaller varieties from 5mm (1⁄10in).

secreted by a gland on the foot, and harden into strong, silken fibres.

Reproduction in the zebra mussel is similar to that of marine mussels. It produces a free-swimming larva which lives on tiny food particles until it matures and settles on a suitable object.

More mussels Other freshwater mussels include twenty species of pea mussel, a tiny but constant source of food for birds and fishes, and the horny orb mussel which is found among water plants in rivers and ditches. The larger freshwater mussels suffer very little from predators. The main threats to their well-being are drought, and the dredging of sluggish rivers and lakes.

Below: The valves of the shell of the painter's mussel were not only used by early Dutch painters as containers for their colours, but were also considered to be an ideal size for the sale of the small quantities of gold and silver leaf used in illuminated manuscripts.

SNAILS OF WET HABITATS

About 15 kinds of snail live in the wetlands of the British Isles. Some are adapted land dwellers, some are aquatic and several are truly amphibious.

Above: Part of a gathering of *Trichia* snails (either *T. plebeia* or *T. hispida*). They have climbed a waterside poplar to escape not flooding but drought–this spot being the last damp place they can reach as their habitat dries out.

Below: A young *Trichia hispida* browsing on a damp patch of leafy liverworts. It is not known for certain whether the hairs on its shell serve any useful purpose.

Marshes and riverside fields provide unstable habitats, fluctuating between land and water. For much of the winter they may be completely submerged, whereas in a hot summer they can dry out with a layer of hard, caked mud. The snails that occupy this intermediate habitat must be able to cope with the variation in conditions, and some are amphibious. The many kinds of snail that are characteristic of wetlands comprise representatives of two distinct evolutionary pathways: those that have adapted to wetland life from a purely aquatic ancestry; and those that originate from snails with fully terrestrial life-styles.

Aquatic ancestry One group of snails, called the Basommatophora because their eyes are set at the base of their tentacles, are primarily aquatic, but at one stage in their evolution they lost their gills and developed the capacity to breathe air from the atmosphere. This allowed them to survive regular exposure on the seashore, or to breathe at the surface when oxygen levels in ponds and rivers became too low. They can also breathe through their skin, while some groups, having reverted to a fully aquatic existence, have developed new forms of gills.

Snails of one basommatophoran family, the Lymnaeidae, can be recognised by their flat, triangular tentacles; most live in fresh water but can often survive buried in the mud of a pond if it should dry up. Some inhabit marshy places, and *Lymnaea palustris* is truly amphibious, crawling freely through damp grass or living a totally aquatic existence in ponds or streams.

Another member of the Basommatophora is the tiny snail *Carychium minimum*, which is a member of the family Ellobiidae. This family is mostly composed of larger snails inhabiting tropical mangrove swamps, but *Carychium minimum* lives in a wide range of damp habitats, being particularly abundant in the flood plains of large rivers.

Eyes on tentacle tips The majority of land snails belong to a group called the Stylommatophora, whose eyes are set at the tips of their tentacles. When the tentacles are withdrawn, a muscle attached to the back of the eyes contracts to turn the tentacles like the inverted finger of a glove. Unlike the Basommatophora, the Stylommatophora are never fully aquatic, but they can inhabit the wet zones intermediate between land and fresh water.

One stylommatophoran family, the Succiniidae, or amber snails, nearly all inhabit the zone extending from the waterside plants to damp meadows. Of our several British species, *Oxyloma pfeifferi* favours the immediate neighbourhood of river banks, whereas *Succinea putris* can be found all over wet meadows.

Some authorities claim that the Succiniidae are a primitive group that demonstrates the transitional stage which land snails passed

through when they evolved from freshwater snails. More recently, the view has been expressed that they are evolved from true land snails but have secondarily acquired features that appear primitive because of their lifestyle in a wet environment.

Rarer snails Some snails that have undoubtedly evolved from fully terrestrial forms now inhabit only very wet conditions. The family Clausiliidae includes members that inhabit a range of conditions, which in southeastern Europe include many species that spend the summer cemented to the hot surfaces of rocks and walls. Most of our British species live in woodland, but one, *Balea biplicata*, lives in rivers in a few widely scattered colonies in the south of England. It was once abundant along the London tidal zone of the River Thames, and continued to survive in colonies at Chelsea, Fulham and Putney until recent times. It is still abundant on Isleworth and Brentford Aits, but these larger Thames islands, which are refuges for a number of rare waterside plants and animals, have been threatened with large-scale development.

On these islands *Balea biplicata* lives under the loose flood debris that lies scattered between the large trees that cover this vanishing type of habitat. A large piece of debris can harbour some tens of this species under it, including numerous young which, because the adults retain the eggs in their bodies until they hatch, are produced fully developed.

Another rare snail that occurs here is a member of the family Helicidae, *Perforatella rubiginosa*. This snail has only been recognised as occurring in Britain in the past few years, although it has been known from parts of Continental Europe since the middle of the last century. *Perforatella rubiginosa* has been overlooked in Britain until now, probably because of its close but superficial resemblance to a number of other members of the Helicidae that live in damp conditions, such as *Trichia plebeia* and *Ashfordia granulata*. All these snails have globular shells that are covered in numerous hair-like growths, and it is easy to see how they have been confused, although internally they are quite distinct.

Perforatella rubiginosa lives in the more muddy areas of Isleworth Ait and Brentwood Ait, and also the intervening stretch of Syon Park that borders the Thames. It has also been found in a similar habitat at Aylesford in Kent but, more interestingly, a smaller form has been found in water meadows bordering the Thames in Oxfordshire and Berkshire.

Although it lives in wet conditions, *Perforatella rubiginosa* will soon drown if immersed in water during the summer. It does not face this risk in winter, for when conditions become cold it survives by burrowing into the substrate, forming a barrier of dried mucus over its aperture, and then hibernating.

Five wetland snails

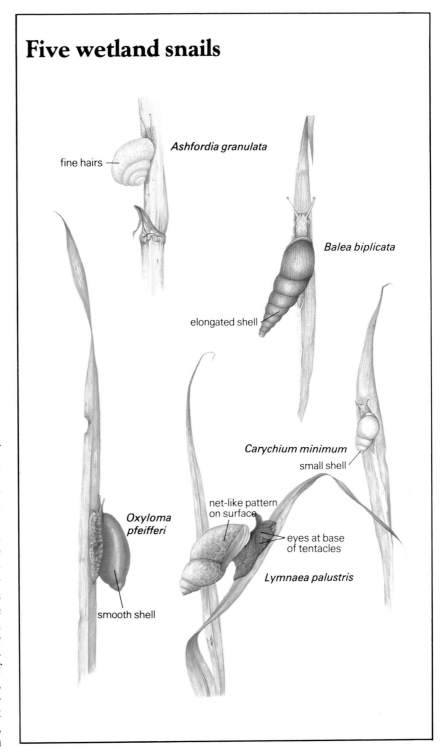

Ashfordia granulata

fine hairs

Balea biplicata

elongated shell

Carychium minimum

small shell

net-like pattern on surface

eyes at base of tentacles

Oxyloma pfeifferi

Lymnaea palustris

smooth shell

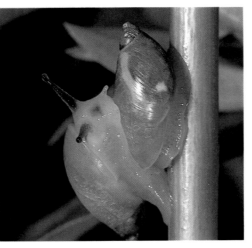

Left: Two amber snails, *Succinea putris*, on a plant stem. They are not aquatic, and might better be called amphibious: these snails can be found either at the water's edge or throughout wet meadows. They climb up tall plant stems like this to escape from flooding, and this is characteristic of the whole of their family, the Succiniidae. It is just possible that this behaviour is an indication that their ancestors may have lived on dry land.

THE FAUNA OF FAST FLOWING STREAMS

Above a certain speed and volume, water carries fine particles away, leaving no foothold for rooted plants. Typical of headstreams, such conditions occur throughout many of our northern river systems. Their stony substratum harbours an abundant fauna.

Above: A fast-flowing Highland river. The animals in such streams face an unending struggle to find enough to eat—and find it before something else does. When animals are many, some face a second problem—avoiding carnivores.

In a habitat where there is plenty to eat you would expect to find plenty of animals. However, if conditions are uniform, as in a stream flowing over smooth rock, or if some non-biological factor is unfavourable, an abundant population may comprise a few species only.

Some animals of fast-flowing streams are specialised to cling to an exposed surface, while others obtain their food by filtering the water, but many show no particular structural adaptations, and find shelter from the current among the stones. In Britain many of these unspecialised animals are confined to running water, where evidently the low temperature and high concentration of oxygen favour their existence; but further north, however, they may extend into lakes. Some animals in our latitudes occur in both still and running water. On the other hand, those that rely on a current to bring their food can live nowhere else.

In a stream, as in other environments, the structure of the community is shaped by the interactions of the species in the struggle to obtain food, and also by the unique feature that the ambient medium, the water, is constantly flowing in one direction—one missed foothold and the individual is swept away.

Composition of the community Insects predominate. The larvae of stoneflies, mayflies and caddisflies are, almost without exception, aquatic. The same holds for two families of the true flies—the non-biting midges (chiro-

Above: A stonefly larva in a hill stream. The food supply is scarce at the top of a stream rising high in the mountains and here stoneflies outnumber all other groups of animals. They appear to be adapted to a meagre food supply and to cold water, for none achieves more than one generation in a year, unlike some other species in different groups. Many grow during the winter and survive the warm season as eggs. The differing levels of temperature and amount of food in a stream thus contribute to faunal diversity.

Below: Blackfly larvae attached to a stone in a fast-flowing stream. They are filter-feeders.

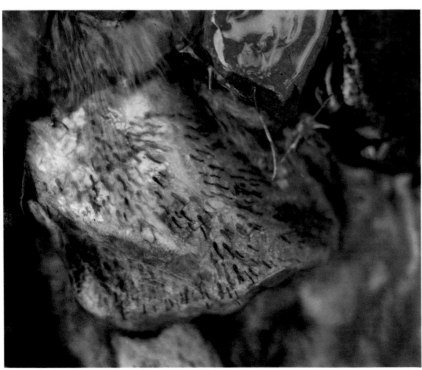

nomids) and the buffalo gnats (simuliids). The winged adults generally live a short life devoted to reproduction and dispersal. Representatives of other insect orders are few.

In other groups the common stream-dwellers comprise a few crustaceans, notably *Gammarus*, the freshwater shrimp, and sometimes crayfishes; a few snails, of which *Ancylus*, the freshwater limpet, is well-adapted to life in a torrent; of arachnids, many species of watermite; the flatworms; some worms; and of the fish, the trout, the bullhead and the stone loach.

Food for all There are two main sources of food for these creatures, the first being, as in most communities, the living vegetation. Algae attach themselves to illuminated surfaces and thrive on the nutrients carried in the continually passing water. The other source, often the more important, comes from the dead leaves of land vegetation. The animals that eat the leaves cannot digest the tissue but fungi invade it and break it down, so providing the nourishment for the leaf-eaters. Large animals with powerful jaws chew and shred the leaves, while the less well-endowed scrape the surface and skeletonise them. Small pieces broken off by the shredders and scrapers, and what has passed through the bodies of these animals, supply food to the detritus feeders.

The algal felt provides food for grazers. Some caddis larvae graze, protected by their sandgrain cases but hindered by them from reaching the most exposed pastures, which are the feeding grounds of two groups of mayflies. The larvae in one group are streamlined and able to swim rapidly; those in the other are flat, with legs spread out sideways to give a broad anchorage so that they can move across the surface with great ability in all directions. Another common grazer in exposed situations is the freshwater limpet which, like its much larger marine counterpart, can resist the current by means of its sucker-like foot and predators by means of its shell. It cannot, however, move fast enough to avoid the rolling stones of an unstable stratum.

Among the stones are cased caddis larvae,

The filter-feeders are adapted to the peculiar conditions of flowing water and can live nowhere else. Larvae of *Simulium* cover an exposed surface with a network of silken strands to which they attach themselves by means of a circle of hooks at the end of the body. They hang from this holdfast and rake fine particles from the current with highly modified mouthparts. Several families of caddis do not make cases but spin nets to filter the water.

The carnivores Several species of stonefly are among the largest of the stream-dwelling invertebrates and they prey on smaller animals. So do the Rhyacophilidae–caddis larvae which make neither net nor case and range freely over the substratum. They, too, are large, and many smaller organisms avoid them by inhabiting crevices too narrow for the hunters to enter. The smallest larvae of some mayflies and stoneflies may be found deep in the substratum.

Current too may offer protection. A herbivore has only to cling and its food lies im-

most of the stoneflies and some mayflies. They feed on detritus and are most numerous where the current is slack, for that is where the detritus settles. Some of the stoneflies have elongated flexible bodies, a modification which enables them to inhabit the narrow passages between small stones.

The main shredders are caddis larvae and the freshwater shrimp, which appears to be intolerant of the low concentration of oxygen found in some ponds, but otherwise has a wide distribution. It is abundant in most streams, notwithstanding a lack of adaptations to life in a current – it is not a good swimmer and has no specialised clinging organs. Its adaptation is one of behaviour; exposed to a current, it immediately seeks shelter among the stones, and it also has a tendency to move upstream.

Above: The dipper is one bird that will actually enter the water to catch invertebrates in upland streams.

Right: The freshwater shrimp *Gammarus pulex*. Pushing and jostling for position among individual shrimps in the stream leads to displacement, producing a fairly even population along the length of the stream.

Below: Most members of the fast-flowing stream community have their own place, to which they are specially adapted.

Life in fast-flowing streams

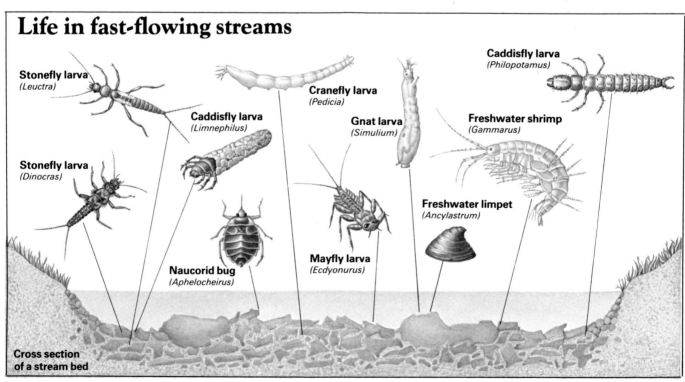

Stonefly larva
(*Leuctra*)

Stonefly larva
(*Dinocras*)

Caddisfly larva
(*Limnephilus*)

Naucorid bug
(*Aphelocheirus*)

Cranefly larva
(*Pedicia*)

Gnat larva
(*Simulium*)

Mayfly larva
(*Ecdyonurus*)

Caddisfly larva
(*Philopotamus*)

Freshwater shrimp
(*Gammarus*)

Freshwater limpet
(*Ancylastrum*)

Cross section
of a stream bed

mobile before it. A carnivore has not only to cling but to stalk, seize and overpower an active prey, which is likely to be difficult in water that flows above a certain speed. The flatworms, another carnivorous group, lacking hard structures and organs with which to seize and overpower prey, entangle it in sticky mucous trails which they lay down on the surface. Trout feed mainly on creatures which have lost their foothold and, of course, at the surface.

Among the birds, the dipper enters the water to catch invertebrates in upland streams. Many other birds enter the food chain and most insectivorous species probably feed, at least occasionally, on the adult insects that emerge from streams – nutritious morsels, for the females are usually heavy with eggs ready for laying. In this way material originating on land, passing to the water via leaves shed in autumn, returns to the land. In no other system is there less re-cycling of organic matter.

Diversity of the fauna The resources are exploited in these various ways, but in each category of exploiters there are many practitioners. It is an accepted ecological principle that two species competing for limited resources must differ in behaviour in some way. One difference common in streams lies in the life history. Among the stoneflies, for example, there are several genera where species emerge one after the other, so that the immature stages are not the same size at the same time and not eating exactly the same food. In other groups the succession is in space, not time, the different species occurring down the length of a water course. A recent study of the Hydropsychidae, a family of net-spinning Trichoptera, has revealed such a succession and shown that each species has a

Above: Warnscale Beck – a typical fast-flowing stream. The effect of human activity on our rivers – especially of our sewers and industrial effluent – has been well documented. The engineer likes vertical sides and a uniform depth, the achievement of which creates less diverse conditions with consequent reduction in diversity of flora and fauna, and power stations raise the temperature of the water. Nor are upland streams immune. Their water, dammed to create a reservoir, is of good quality and gravity takes it to where it is wanted. The resulting changes in flow, temperature and food supply affect the natural fauna. Also, in the uplands there are deposits of copper, lead and other metals. Few are exploited today, but the spoil heaps remain and rain leaches the metals from them in harmful concentrations.

Left: A bullhead camouflaged on river bed gravel. Trout and stone loach can also be found in upland streams.

different optimum temperature, coupled with which is an increasing tolerance of low oxygen concentration.

Other changes in the environment also make for faunal diversity. A dam, for instance, makes conditions suitable for plankton, some of which, passing downstream, increase the food for net-spinners and lead to a rise in their numbers.

Colonizing the length of a stream A net set in the open water gathers a varied selection of the inhabitants, particularly during the night. Species tend to drift downstream when they are growing most rapidly and therefore, presumably, feeding most avidly, and this occurs at times when there is no increase in flow.

No-one has observed exactly what happens, but probably attempts to reach ungrazed pasture where the current is just too strong lead to dislodgement, and there may be pushing and jostling in a dense population. Many individuals are not carried far before they regain the safety of the stones, but there is a constant displacement downstream of most of the fauna. Obviously *Gammarus*, aquatic throughout life, has no option but to walk back to reach the ungrazed pastures and its tendency to move upstream leads to overcrowding at the top, increased pushing and jostling and increased displacement, which eventually produces a uniform population throughout the length of a stream.

INSECTS OF THE WATERY JUNGLE

Beneath the surface of our freshwater rivers, streams and lakes exists a watery jungle in which predatory insects stalk their herbivore cousins.

A remarkable variety of insects have become adapted to life in the water or on its surface, some browsing on aquatic plants while others are predators on the browsers. The vegetarian insects are often present in great numbers, far more than their predatory adversaries. However, it is important to note that the distinction between predators and

Above: Many insects of ponds and streams pass their adult stage out of water. One such is the common or blue tailed damselfly (*Ischnura elegans*), the adult of which bears no resemblance to its dull brown water-dwelling nymph. The airborne stage of the damselfly's life allows the adult to spread its eggs to other rivers and streams.

Right: Among the insects that pass their adult, as well as larval stage in water is the great diving beetle (*Dytiscus marginalis*). A good swimmer, aided by its powerful paddle-like hindlegs, this species is one of the largest and most ferocious of aquatic insects.

vegetarians is often blurred among aquatic insects: for example, the larva of one species might be a predator while the adult is a vegetarian, or *vice versa*.

The struggle for life in water involves different insects in four main habitats within the pond or stream: the surface, just below the surface, the open water and the bed of the pond or stream. Each habitat has its own community of predators and vegetarians, each adapted to the particular problems of its chosen niche.

Surface life The surface of the water collects floating debris, such as pollen grains, dead insects and leaves, that provides food for a variety of insects. This plant material attracts large congregations of the most primitive of insects–springtails. There are two British species which have adapted to life on the water, the most common of the two being the tiny (1.5mm)* *Podura aquatica*–seen with the naked eye as blue-black clusters of creatures which leap about on the surface of stagnant waters when disturbed.

Along with the bodies of dead insects these springtails are preyed upon by a whole host of bugs and beetles. These include the water-crickets and pond skaters, both of which belong to the bug order (Hemiptera), despite their common names. As with all bugs these surface dwellers have piercing mouthparts with which they suck the juices of their prey.

A common surface predator is the whirligig beetle which feeds upon small insects that fall on to the water surface. Unlike the predatory bugs, the whirligig beetle can dive for food beneath the water surface.

Under the film Living just below the surface film of still water are the vegetarian larvae of some species of biting midges and non-biting midges. These feed on algae and decaying

Encounters in the water

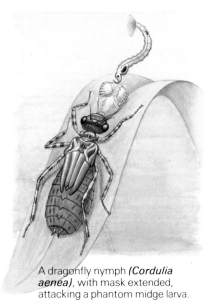

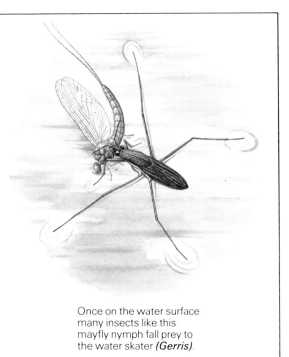

A great diving beetle larva *(Dytiscus marginalis)* capturing a mosquito larva.

A dragonfly nymph *(Cordulia aenea)*, with mask extended, attacking a phantom midge larva.

Once on the water surface many insects like this mayfly nymph fall prey to the water skater *(Gerris)*.

plant material, respectively. Living plants also provide food for some aquatic beetle larvae and at least one group of moth caterpillars called china-mark moths. The larvae of reed beetles–long, narrow beetles with brilliant metallic sheens–feed on submerged plants and obtain their oxygen supplies by tapping into the air spaces of their foodplant. The caterpillars of china-mark moths live inside protective cases made from fragments of the floating leaves of water plants beneath which they feed.

Life is hazardous close to the surface, as many of the larger insect predators regularly rise to the surface to replenish their air supply, and may take a quick meal at the same time. Species such as the water boatman and the great diving beetle would find easy pickings among the herbivorous larvae that they see on the leaves of surface plants.

Open water Many of the insect predators are found swimming in the areas of open water. Among the more prominent are the water beetles; these include some of the largest aquatic insects. Among these are the fearsome *Dytiscus* species, commonest and largest of which is the great diving beetle (*Dytiscus marginalis*) with a length of 3-5cm (1½in). It attacks other aquatic creatures, including small fish and tadpoles as well as virtually any other insect it encounters.

Equally ferocious are the greater water boatmen of which we have four species in Britain. These large bugs, some 15mm long, often take on insects, and even fishes, bigger than themselves. They subdue their prey by injecting them with a poison.

Other predators of open water include the smaller beetles such as *Hydrobius fuscipes* and *Acilius sulcatus*, and species belonging to the genera *Agabus* and *Hydroporus*. All these are much smaller than the great diving beetles and

Right: The predatory water scorpion (*Nepa cinerea*) owes its name to its large grasping forelegs and its long breathing siphon, which is often mistaken for a sting.

Below: Another common bug of streams and ponds is the greater water boatman (*Notonecta glauca*). Like the water scorpion, it has a sharp, pointed proboscis which it uses to suck prey dry of body fluids.

Above: Dragonfly nymphs have an unusual mask-like set of jaws hinged to the underside of their head. These can be shot out with amazing speed to capture suitable victims, such as insects, small fishes, tadpoles or (above right) young newts.

Left: The water stick insect (*Ranatra linearis*) resembles a thin version of the water scorpion–it has the same grasping forelegs and long breathing siphon.

Below: The great silver water beetle (*Hydrophilus piceus*) is mainly vegetarian in the adult state yet a carnivore as a larva.

take proportionately smaller prey. A popular item of prey for these beetles are the larvae of flies such as non-biting midges and phantom midges. The adult screech beetle is one of the more interesting species which seek out other insect larvae in open water. It owes its name to the ability of the adult beetle to squeak loudly when disturbed.

Life on the bed By far the most prolific place to live for many vegetarians and predators alike is the bed of the stream or pond. Here are found a whole range of water plants and a rich supply of organic silt and debris to form a basis for a food chain.

This is where the larval, or nymphal, stages of most aquatic insects live and feed. All face the problem of how to obtain oxygen without continually moving up to the surface to replenish stores. Most overcome this problem by possessing gills which allow oxygen from the water to diffuse into the insect. However, not all species live in oxygen- rich water; some larvae, especially those of flies, live in muddy deoxygenated places in stagnant ponds and ditches. The larvae of some species of non-biting midges overcome this problem by extracting oxygen from their surroundings and then storing it by means of haemoglobin in their bodies. The so called rat-tailed maggots–larvae of hoverflies such as *Eristalis* or *Helophilus*–have snorkel-like siphons which can extend up to 15cm (6in) to reach the water surface and air.

Although some species of mayfly and stonefly nymphs are carnivorous, most are vegetarian and eat fragments of plants. The mayfly nymphs may live anything from a few months to three years, depending upon the species, before reaching adulthood. Stonefly nymphs, also known as 'creepers', especially to fishermen, live for one to three years before becoming adult. Nymphs of these two insect groups are often abundant in streams and ponds and provide a rich source of food for other insects, fish and even birds such as the dipper.

Every stretch of standing or running water holds at least one species of caddisfly whose larvae usually conceal themselves in protective cases of sticks, gravel, leaves or even snail shells. Most species browse on living and dead plant material although some, such as *Hydropsyche angustipennis*, do not build cases but trap small animals in a net. To do this they pick a stone in a fast-flowing stream and underneath it they spin a web of silk with the opening facing upstream. This behaviour is typical of aquatic insect groups–some are vegetarian while others specialise in eating smaller creatures, insect or otherwise.

Aquatic tigers The most abundant of bed and plant dwelling insect predators are the larvae of water beetles. Just as the adult beetles rule the open waters, so the larval stages are the 'tigers' of the pond-bed. Even species such as the great silver water beetle whose adult is a vegetarian have predatory larvae with sharp curving jaws with which they feed on water snails. But the most ferocious of all the water beetle larvae is that of the great diving beetle, which may grow to 5cm (2in) in length. With their powerful jaws these larvae can overpower other animals as large, or even larger than themselves. Some water beetle larvae appear to specialise in one sort of prey–that of the great silver water beetle seems to prefer water snails, while the larva of the screech beetle feeds almost exclusively upon *Tubifex* worms which it seeks in the mud.

Patient bugs In contrast to the energetic water boatmen or pond skaters many bugs use the strategy of waiting for a meal to come to them. The water scorpion (*Nepa cinerea*) can grow to a length of 3cm (1¼in) and has a long spiny 'tail' through which it obtains oxygen by pushing it above the water surface. These bugs sit among water plants and rely upon their mottled brown camouflage to blend into the background. Any suitable sized insect, fish or other small creature that comes

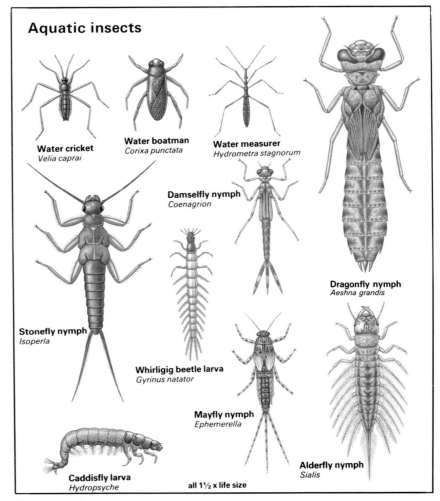

Aquatic insects

Water cricket
Velia caprai

Water boatman
Corixa punctata

Water measurer
Hydrometra stagnorum

Damselfly nymph
Coenagrion

Dragonfly nymph
Aeshna grandis

Stonefly nymph
Isoperla

Whirligig beetle larva
Gyrinus natator

Mayfly nymph
Ephemerella

Alderfly nymph
Sialis

Caddisfly larva
Hydropsyche

all 1½ x life size

Below: An abundant source of food for predators in summer is mosquito larvae, which hang down from the underside of the water surface. These larvae have vibrating bristles around their mouths to trap food.

within striking distance of the water scorpion's menacing forelegs is grabbed and sucked dry.

A similar but longer and more slender bug which uses the same strategy is the water stick insect (*Ranatra linearis*). Unlike the water scorpion this bug is not totally reliant upon stealth as it can swim well and is capable of flight.

About 6.5cm (2½in) in length, the water stick insect has long thin legs. The forelegs are used for grasping their prey, which consists of much the same species preyed on by the water scorpion.

Beauty and the beast Rivalling the great diving beetles as the tigers of the watery jungle are the nymphs of the dragonflies and their smaller relations the damselflies. These are both stalkers which rely upon stealth and patience to grab suitable prey in their extendible jaws or 'mask'. Once caught in the pincer-like grip the prey is pulled back under the nymph's head and eaten.

The adult and nymphal dragonflies not only exploit different food sources, just as their land-dwelling counterparts do, but they also live in different environments. This is an important strategy allowing insects to diversify and colonize a wide range of habitats. Few insects show this better than the dragonflies – the contrast between the ugly functional nymph and the sleek sparkling adult is almost unparalled in the insect world.

CASE-BUILDING CADDISFLIES

Caddisflies, small drab brown insects, are famous for the ability of their larvae to build protective cases out of bits of twigs, leaves or gravel, inside which they can survive the unwelcome attentions of a multitude of predators.

Top right: A cluster of newly emerged grannom caddisflies (*Brachycentrus subnubilus*) caught on the sticky hairs of comfrey on the bank of a river.

Below: The larval case of *Phryganea grandis* is made up of little bits of vegetation carefully cemented together with silk.

Caddisflies, often known to anglers as sedges or rails, are common insects of rivers, streams, lakes and ponds throughout the British Isles. There are nearly 200 different species. They belong to the insect order Trichoptera, which means 'hairy wings' and refers to the minute hairs that cover the veins and membranes of each of the two pairs of wings. One other noticeable feature is the long antennae which are held straight out in front of the head when the insects are at rest. These antennae are frequently as long – if not longer – than the wings.

Most caddisflies are rather difficult to see; they flit about at dusk and can easily be mistaken for small, brownish moths. In the daytime they hide among waterside vegetation and rarely feed. Small species usually stay close to their home stream or pond, but larger ones – their powers of flight being stronger – venture farther afield.

Caddisflies are an important link in the freshwater food chain since very many fishes and water birds feed on both adults and larvae. Female caddisflies laying their eggs near or over water are particularly attractive to fishes – this is why so many anglers model their artificial flies on caddis species.

Self-protection in a watery world The fully aquatic larval stage of caddisflies is much better known than the adult stage, because the larvae are frequently dredged up out of ponds and streams by children with jam jars looking for frog spawn, beetles and other freshwater creatures.

The larvae have soft, caterpillar-like bodies – making them easy meals for predators – so protection of some kind is vital. The unique answer the caddis larvae have come up with is to build protective cases for themselves, in which they spend their entire larval and pupal lives. Species living in still and slow-moving water build cases out of grains of sand, bits of twigs and leaves, bark or even tiny mollusc shells. Those in fast-flowing streams weave silk nets for protection.

There are as many as eight different methods employed by case-building larvae to make their cases, but in all examples the building materials are cemented together with sticky silk produced by a special gland near the mouth. Each species favours different materials and makes a case of a recognisable pattern – so much so that it is often possible to identify the species by looking at the case alone. The larva winds the silk around itself and then attaches the building materials to this 'sheath'.

The larva, which has a fairly hard head and thorax, attaches itself firmly to the inside of the case with a pair of hooks situated at the tip of the soft abdomen. As it grows, it adds more material to the front end of the case. The larva breathes underwater by means of feathery gills, undulating its body within the case so that a current of water passes over the gills. The cases are always open at both ends, though the rear opening may be very narrow. The head and front legs protrude from the case so the larva can move in search of food. With their protective 'houses' the larvae can live in streams where trout or other fishes live. (Sometimes, however, fishes feed avidly on caddis larvae, having no trouble in swallowing the case as well as its occupant.)

Case-building caddis larvae are all vegetarian feeders; net-makers, on the other hand, are generally carnivorous. They spin non-movable silken nets and attach them to the underside of a stone, or on submerged vegetation. The larvae of the Hydropsyche family, common in fast-flowing streams, position the net so that its open end faces upstream. Small animals washed down by the current are snapped up by the larva sitting in the net. Members of the *Rhyacophila* genus are different again, being free-living, tougher-bodied predators.

Pupation The larval stage lasts about a year. Most species pupate in the spring, emerging as adults in the summer, but a few overwinter as pupae. When the time comes for pupation, the larva seals itself up inside the case and cements it firmly to a submerged object. Non case-building larvae pupate in a silk cocoon inside a specially made chamber of sand.

The caddis pupa is fairly active and continues the undulating movements necessary for the circulation of oxygen-bearing water. It has a large pair of jaws with which, when it is ready, it bites its way out of the case and crawls or swims actively to the surface, where the adult finally emerges and almost immediately takes to the wing.

Adult life in the air If the larval caddisfly has been fortunate enough to escape the predatory attentions of fishes and water birds, it still has to face a multitude of other enemies as an adult. Bats and birds, in particular, take a heavy toll.

The newly emerged caddisflies fly in large numbers over the water, often gathering together in little swarms. Male and female mate while resting on a plant. The eggs are laid by the female in a gelatinous mass – either on overhanging leaves or, in running water species, on stones or vegetation under water. The eggs hatch in two or three weeks, those larvae which are out of water finding their way into it simply by falling in. The adults do not survive long after mating and egg-laying; their lifespan is generally only a few weeks.

Some species The largest caddis larva is that of *Phryganea grandis*, which lives in still or very slow-moving water and builds cases out of spirally arranged plant material. The adult insect has a wing-span of about 5cm (2in). Another still or slow-moving water species is *Limnephilus rhombicus*, which makes a case of small mollusc shells. A member of the same family as *Limnephilus*, *Anabolia nervosa* makes a case of sand grains to which are attached several twigs – the whole a device to make swallowing by a fish extremely difficult.

An unusual species is the grannom (*Brachycentrus subnubilus*), which fixes its case – made of bits of leaves and stems – to the vegetation and filters food from the water with its comb-like middle legs. *Hydropsyche* species are unusual, too, in that the adults can be seen flying in bright sunlight. These are species of streams and rivers.

Above: Adult caddis fly; it holds its two pairs of wings above the body when at rest. Note long, slender antennae.

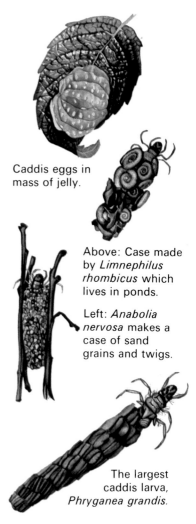

Caddis eggs in mass of jelly.

Above: Case made by *Limnephilus rhombicus* which lives in ponds.

Left: *Anabolia nervosa* makes a case of sand grains and twigs.

The largest caddis larva, *Phryganea grandis*.

Larva of *Hydropsyche angustipennis* lives upside down in net.

Left: Caddis larva with case of twigs and plant debris; lives in slow-moving water and drags case with it as it looks for food.

Above: Free-living larva of *Rhyacophila septentrionis* preys on small water animals.

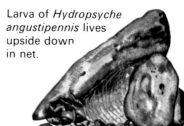

DELICATE DAMSELFLIES WITH GAUZY WINGS

Damselflies, resembling small vividly coloured dragonflies, are usually found on the wing in summer, when they can be seen flying by rivers, slow streams and canals. They are smaller than the dragonflies, but have the same net-like veining on the wings, and large, bulging eyes.

The insect order Odonata, the dragonflies, is represented in Britain by two suborders, the Zygoptera, or damselflies, and the Anisoptera, the dragonflies themselves.

There are several basic differences which enable damselflies to be distinguished from dragonflies. The latter are large insects with stout bodies and powerful flight. Their hind wings are broader at the base than their forewings and, when at rest, the wings are always held out flat on either side.

Damselflies are smaller insects with slender bodies and weak, fluttering flight. Their fore and hind wings are similar in shape and, when

at rest, are held over the back, either together or slightly parted. Their eyes are more widely separated than those of dragonflies.

However, the two groups also have several features in common, including the fine, net-like veining of the wings, the large eyes and very small antennae, the predatory habits of both the aquatic nymphs and the adult stages, and the distinctive method of mating.

There are 17 species of damselfly listed as British, but two of these have not been seen since the 1950s and are probably now extinct in Britain. Despite the relatively small number of species, damselflies are not at all easy to

Above: The male of the large damselfly, the banded agrion (*Calopteryx splendens*), is unmistakable, due to the broad, iridescent purple-brown or deep blue band on its wings. This appears near the tip of all four wings, the rest of the wing being only slightly tinted. The female (shown above left) is not banded and her wings have a rather brownish tinge. Damselflies rest in a position more akin to butterflies than to dragonflies, with their wings held together over their backs.

identify. Almost half the species have blue and black bodies, with very variable markings. Also, the full colour pattern does not develop until a few hours, or even days, after emergence, and the early stages of the damselfly are comparatively dull and pale.

Mating and reproduction The process of courtship and mating is very similar in dragonflies and damselflies, but differs from that of other insects. The genital opening of both sexes of damselfly is near the tip of the tail, but before mating the male transfers his sperm to an accessory genital organ on the underside of his abdomen, just behind the thorax. He then finds a female and seizes her by the back of the neck with a pair of claspers situated at the hind end of his body. They then fly about in tandem, and settle, linked together in this way. When mating takes place the female bends her body round under the male's body, and the sperm is transferred for fertilisation.

In some species, a form of courtship has been observed before mating. The male banded agrion, for example, signals to a passing female by bending his abdomen upwards and spreading his wings. If the female responds by landing near him he performs a fluttering, aerial dance over her before settling and taking up the tandem position.

Egg-laying The female damselflies lay their eggs under the water, some inserting them in slits in the stems of plants which are cut with a saw-like ovipositor situated at the rear end. The male and female often remain in tandem during egg-laying, and sometimes both will walk down the stem of a plant until they are submerged to a depth of several inches. They may stay in this position for 10 or 15 minutes, a film of air clinging to the thorax of both damselflies and giving them a silvery appearance. When they emerge again they need only a few minutes to dry their wings before they are able to fly away.

Above: The male demoiselle agrion (*Calopteryx virgo*) has rather spectacular colouring. The body is a dark, shining green and the wings are deep purple-brown to dark blue, becoming bluer with age. This newly emerged male has not yet developed the final deep coloration on its wings.

Below: The red-eyed male of the large red damselfly (*Pyrrhosoma nymphula*), eating a small fly. Damselflies feed on other insects, using their large, efficient eyes to find and pursue them. They may catch their prey on the wing, or take resting insects from reeds or herbage. In either case, the front legs, which are directed forwards, are used to capture and hold the victim.

The eggs hatch to produce larvae known as nymphs because of the absence of a pupal stage. The nymphs spend up to two years underwater, depending on the species, living at the bottom among the aquatic plants. They climb out of the water for their final moult, when the winged damselflies emerge.

Two large damselflies The demoiselle agrion and the banded agrion are our largest and most beautiful species of damselfly. The demoiselle agrion is found mainly in southern England, particularly in the New Forest and in Cornwall, and also in Ireland. The nymphs live in fast-running, clear streams with a pebble bottom. The metallic green body of the adult measures about 4.5cm (1¾in) long, and the male has dark, iridescent wings. The wings of the female are dull brown, with distinct dark veins, and a white spot appears near the tip of each wing.

The banded agrion occurs all over southern England and in Ireland, and may be abundant in favourable localities. The nymphs are found in fast-flowing streams with a sandy or muddy bottom. The adult's body is dark, metallic green and the male has distinctive bands across all four wings. The female is not

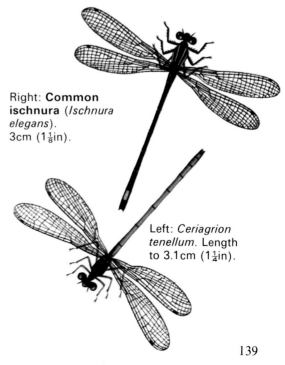

Right: **Common ischnura** (*Ischnura elegans*). 3cm (1⅛in).

Left: *Ceriagrion tenellum*. Length to 3.1cm (1¼in).

banded and resembles the female demoiselle agrion.

Five colourful species The large red damselfly first appears in mid-May, rather earlier than the other species. It breeds in lakes, canals and slow-moving, clean, fresh water and is one of our commonest species, being found all over the British Isles.

Both sexes have a red abdomen with black markings, a black thorax striped with red, and black legs. In a rare variety of the female, the thorax is striped with yellow and the abdomen is black with narrow yellow rings. The overall body length is about 3.6cm (under 1½in); the other red species, *Ceriagrion tenellum*, is smaller and has pale red legs.

The common coenagrion is found in England and Ireland (but not in Scotland) wherever there are clean ponds, ditches or canals with plenty of reeds and long grass. It is the commonest of several species in which the male is predominantly blue, marked with black. The male coenagrion is distinguished by a U-shaped black mark on the upper side of the second abdominal segment. The female is normally black, with a little blue at the tip of the body, but sometimes resembles the male. Both sexes measure about

Above: A mating pair of the common coenagrion (*Coenagrion puella*). The female–the back of her neck held by the male's claspers–bends her body round, underneath that of the male. Her genital opening makes contact with the male's accessory organ and sperm is transferred to fertilise the eggs.

Opposite page: A pair of demoiselle agrion damselflies. The male is on the right, the female left.

Below: The nymph of a damselfly does not pupate. After several moults, when it is almost mature, the nymph–like this one of the common coenagrion–climbs out of the water on to a reed, the skin splits open and the winged damselfly emerges. The full adult coloration takes a little while to develop after emergence.

3.3cm (1¼in).

The common ischnura is widespread in ponds and ditches and seems to be more tolerant of pollution than the other species. Both sexes are black, with a blue band near the end of the body which may be absent in the female.

The green lestes (*Lestes sponsa*) which is found throughout the British Isles, lives and breeds in open ditches, ponds and lakes fringed with rushes or reeds. Measuring 3.8cm (1½in) it has a metallic green body and clear wings.

Damselfly rarities Two less common species are easily recognised and worth looking out for; both are local in southern England. The white-legged damselfly (*Platynemis pennipes*) is a pale coloured insect with white, feather-like hind legs. The red-eyed damselfly (*Erythromma najas*) is similar in coloration to the common ischnura, but is more robust in appearance and has bright red eyes. If you find one, photograph it, don't collect it.

Water pollution is the major threat to damselflies. A site in the Norfolk Broads, visited in 1975, was found to contain only one species–the common ischnura–whereas 30 years previously 16 species of damselflies and dragonflies were living there, including one species now extinct. Unless the pollution of such sites is controlled, then a further rapid reduction in numbers of these beautiful aquatic insects seems more than likely.

Larval lifestyle

Damselfly larvae or nymphs differ from those of dragonflies in being more delicate and slender (like the adult insects). They also have three conspicuous, leaf-like gills at the hind end of the body, through which they breathe, whereas the gills of dragonfly nymphs are concealed internally. The larvae spend one to two years underwater, living at the bottom, among aquatic plants. They are highly carnivorous, feeding on a variety of small aquatic insects and crustaceans.

All Odonata larvae capture their prey by means of a greatly enlarged labium, or lower lip. This is jointed so that it can be extended forwards or withdrawn under the head, and it has a pair of grasping claws at its tip.

When suitable prey is spotted, it is stalked to within a few millimetres and the labium is shot out, then pulled back with the victim firmly in its grasp. The labium is not projected by direct muscle action, but by an increase in blood pressure following the contraction of muscles in the abdomen.

dragonfly larva   **damselfly larva**
gills concealed   conspicuous gills

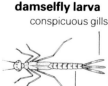

slender body

Above: The croaking of the male frog in the breeding season is a guiding call to females in the area, and helps direct them to the spawning site. It also signals the preliminary stages of courtship with the female, which only grunts in response.

Above: When the frog is swimming its pale underside camouflages it against light filtering down through the water. If the frog had a dark belly it would be outlined against the light and therefore easily spotted by underwater predators.

THE AGILE COMMON FROG

Famed for its leaping ability (powered by its long, muscular back legs), the common frog is easiest to spot in spring when the adults emerge from their winter retreats.

The common frog is present throughout Britain and Ireland, except for the Outer Hebrides and Orkney, and has an extensive range of habitat–from sea level to nearly as high as the snow line on mountains 760m (2500ft) up. Its usual habitat is damp vegetation and near water and it is often a visitor to suburban gardens.

The frog is not the easiest creature to spot since it spends most of its time sitting among thick vegetation, where it is well camouflaged. Your best chance of seeing one is by disturbing it–although you will have to be on the alert, because at the first sign of danger it is off with a couple of startled leaps into nearby foliage.

Colour and markings The shade and marking of the common frog's skin vary enormously. The basic colour ranges from a pale greenish-grey, through bright yellow to a dark olive-coloured brown. The skin can be marked with spots, speckles or marbling in black, brown or red; in Scotland large, reddish-coloured frogs are quite common. The only regular markings are the dark cross bars on the limbs, and streaks behind and in front of the eyes. When the frog crouches the dark bars on the thighs, calves and feet are aligned to form continuous dark streaks. These and other patterns break up the frog's outline and help it to merge into the background.

The frog can also change its colour, lightening or darkening its skin by contracting or spreading dark pigment cells scattered over its body beneath the outer skin. The basic pattern remains the same, but within an hour the frog can assume a completely different colour. In lighter, warmer and drier conditions it becomes paler, but darkens when in colder and damper weather.

The skin is smooth in texture with numerous small bumps on the flanks. The only exception occurs when the female develops a rough skin texture in the spawning season, which enables the male to distinguish the sex. Special mucous glands in the skin keep it moist, so the frog can breathe through its skin and supplement the respiration of its

Left: The frog's large protruding eyes are a distinctive feature. The pupils can contract to a horizontal slit in bright light, but the eyes can only be closed completely when the frog withdraws the eyeballs into its head, which it does when eating.

Catching prey
the tongue is joined to the front of the mouth

The frog catches moving prey by flicking out its long tongue. When the food is inside the frog's mouth, it is squashed between the tongue and the eyeballs (which are drawn down inside the head).

Below: Common or grass frog *(Rana temporaria)* meets common toad *(Bufo bufo):* the frog is generally slighter, with a maximum size of up to 10cm (toads up to 15cm). Its skin is smoother than that of the more warty toad. In the mating season a male frog may even jump on to a toad's back in the mistaken belief that he has found an ideal mate.

simple lungs. The skin can also absorb water, so the frog does not need to drink. As it grows, the frog regularly sheds a transparent surface layer of dead skin (because dead skin cannot stretch as the frog grows). Just before this happens, the mucous glands become very active, lubricating the new skin underneath. When the surface of the skin starts to split, the frog uses its feet to push off the old covering, which it generally eats afterwards.

Eyes and mouth One of the most distinctive features of the frog is its large jewel-like eyes. The glistening iris is brown, flecked with gold, and the limpid, large round pupil contracts in bright light to a horizontal slit. Each protruding eye is protected by a thick immovable upper and lower lid and a thin movable transparent inner eyelid which is known as the nictitating membrane. This can be raised from beneath the lower lid to cover the eye, especially when the frog is underwater. The only way the frog can close its eyes so that the upper and lower lids meet is to withdraw the eyeballs into its head. It does this when swallowing food, gulping and blinking at the same time; pressure from the back of the eyeballs helps force food down the frog's gullet.

The frog makes good use of its wide mouth and long tongue to snap up whole invertebrates. Slugs and worms are a favourite diet, but the frog also catches flies and insects which might be expected to escape such a sedentary creature. The free end of the tongue, which points down the throat, can be projected with a whiplash action at great speed to snatch unsuspecting prey nearby. The frog has numerous, minute cone teeth around the edge of its jaw and two patches of teeth–vomerine teeth–in the roof of its mouth; these prevent slippery slugs, snails and worms from escaping once caught in the frog's mouth.

Breeding cycle Frogs are not usually seen

until February or March, when the adults emerge from their winter retreats–ponds, ditches and occasionally on dry land. They now begin to congregate at various breeding sites, preferring ponds that have water flowing in and out of them, and canals. Frogs may travel a distance of half a mile to reach their spawning site, where they gather together in large numbers particularly on mild rainy nights–an amazing spectacle. The males, who always arrive first, strike up a croaking chorus–a 'grook-grook-grook' call –to attract the females. They produce this mating call by closing the mouth and nostrils firmly and gulping air backwards and forwards over the vocal chords. Frogs can amplify their croaking by puffing out their throat pouch with the internal vocal sac. The female frog only utters the odd grunt; she is never as vociferous as the male.

Free-for-all mating Frogs do not display any elegant courtship rituals; the eager male simply grabs the nearest female as she arrives at the spawning site. So strong is the sexual urge that the male in his frenzy may even grab another male, a fish, a stick or, if offered, a human finger. Jumping on the female's back, the male wraps his forelimbs around her body

Metamorphosis . . . from egg to frog

Each frog's egg, 2-3mm in diameter, is enclosed in an envelope of jelly. When the egg is deposited into water the jelly swells to a diameter of 8-10mm, insulating the eggs from the water. The egg develops into a tadpole in 10-21 days (the higher the temperature the shorter the time). The tadpole digests the jelly using a secretion from a special gland, and adhesive organs help attach the tadpole to other spawn or water plants. Until the mouth forms, the tadpole gets nourishment from the remains of the egg yolk. It then starts to eat algae, breathing by means of three pairs of external gills, which are soon covered by a flap of skin. An internal gill cavity is now used for breathing, connected to the outside by a small hole (spiracle) in the left side. Normally hind leg stumps appear after five weeks; by the seventh week toes have formed. At eight weeks lungs have developed and the tadpole surfaces to gulp air. By the twelfth week development is accelerating and the forelimbs are visible. The spiral intestine is shortening in readiness for a carnivorous diet, while the substance of its tail is transferred to the body by a process known as resorption which adds nourishment for more growth. The final stage takes ten weeks and by May/June the young frog, now 12-15mm long, spends most of its time on rocks out of water or in nearby damp grass. Scarcity of food or cold conditions may delay metamorphosis and overwintering tadpoles are not uncommon in the north. Young frogs are 20mm long by October/November and double in size by the following autumn. They reach sexual maturity in the third year.

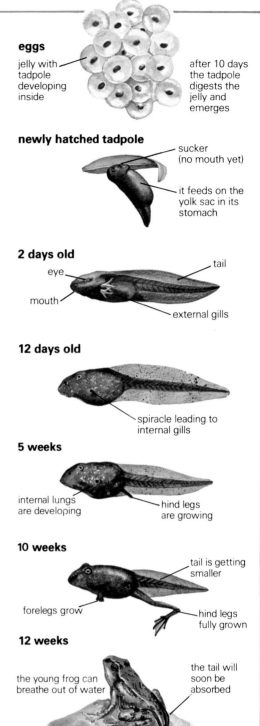

eggs

jelly with tadpole developing inside

after 10 days the tadpole digests the jelly and emerges

newly hatched tadpole

sucker (no mouth yet)

it feeds on the yolk sac in its stomach

2 days old

eye

mouth

tail

external gills

12 days old

spiracle leading to internal gills

5 weeks

internal lungs are developing

hind legs are growing

10 weeks

tail is getting smaller

forelegs grow

hind legs fully grown

12 weeks

the young frog can breathe out of water

the tail will soon be absorbed

Male or female?

In the breeding season these two swellings (nuptial pads) on the male frog's forelimbs (above left) are covered with a dark, rough layer of skin.

Left: A common frog sitting on a water-lily leaf.

just below the 'armpits' and grips using his nuptial pads—a position called amplexus.

The spawning itself, which takes place in water, can happen at any time during amplexus and lasts just a few seconds. The female lays over 2000 black eggs by pressing her forelimbs on her abdomen as the male releases sperm. The eggs are fertilised by the male's sperm immediately they are laid, and before their gelatinous capsules absorb water and swell up. The capsules swell very soon and this reduces their density. So the spawn, which at first sinks to the bottom, floats to the surface, joining up with all the other spawn in one mass.

After spawning, the female normally leaves the pond while the male often goes in search of another mate. Both male and female frogs return to the same breeding site year after year, probably recognising it by the smell of the water and algae.

Survival Only a few of the masses of tiny frogs survive to adulthood. Most perish or are taken by predators such as herons, gulls, ducks, snakes, hedgehogs, shrews, badgers, rats, weasels, stoats, otters, mink and foxes. Drainage of wetlands, the constant dredging of rivers and the decrease of farm ponds have meant fewer suitable places in which to live. Pollution of the water and the indiscriminate use of insecticides have also affected the frogs' habitat.

Birds of river and stream

The majority of breeding birds found using our rivers, streams and canals depend not only on the linear watercourses themselves but also to a vital extent on the mosaic of habitats which flank them. One without the other is rarely good enough.

A walk along open stretches of a river or canal interspersed with sections bordered by reed beds or woodland and scrub cannot fail to instil the importance of these hinterland areas for the birds of our waterways. In the open areas your eyes may be drawn to a stately family of mute swans or a young brood of mallard, while among the reeds and thickets the calls of buntings, whitethroats, tits, warblers and a host of others sound from all sides. You may also be rewarded by the sight of a silent, shimmering blue dart as a kingfisher disappears, rapier-like, downstream.

Upland rivers, which tumble over boulders and have rocks instead of emergent vegetation flanking their sides, attract different birds altogether. You can expect to see the bobbing dipper, but the grey wagtail is equally at home here. The two contrast markedly, however: the dumpy dipper swallows stones to weigh itself down as it feeds submerged in the torrent, while the wagtail flits over stones like a ballerina, snapping up insects as it goes.

The differences between a canal and a lowland river can be quite profound for a bird. Although most prefer the undisturbed tranquillity of a river miles from the beaten track, the threats to raising a family here may be greater from sudden flash floods than from the seasonal hustle and bustle of the canal with its stable water level. The answer for the birds is to nest and raise a brood quickly and move off before the peak holiday rush brings picnickers, campers, fishermen and boating parties. For birds such as the coot, great crested grebe and dabchick, this may mean raising a 'canal' family in spring and a 'river' family in the summer. In this way, success can be assured.

Left: All you are likely to see of the kingfisher is a sudden brilliant flash of turquoise blue and orange as it darts along a stream. The nest site of this beautiful bird is not so attractive: it consists of a hole in the bank leading to a tunnel, at the end of which live the chicks in a safe — if dark and extremely smelly — chamber.

Left: A grey heron at its tree-top nest. You may spot a heron standing motionless, possibly on one leg, on the bank of a river, watching for a fish to appear. When the prey is sighted, the heron stabs at it with its long beak, then rises into the air to carry the victim back to the nest.

Downy cygnets take to the water early and are guarded carefully by their parents, who pull up underwater vegetation for them. It has been suggested that the adults also paddle vigorously to bring food to the surface for the young.

SWANS: ARISTOCRATS OF THE WATERWAYS

The mute swan that once graced the tables of medieval banquets is now protected from this fate. It is, however, vulnerable to the modern hazards of overhead power cables and lead weights in fishing tackle. Despite its name, the mute swan will snort and hiss noisily if it feels threatened.

Mute swan (*Cygnus olor*)
Length (bill-tip to tail-tip)
152cm (5ft); weight—cob
(male) 12-20kg (26-44lb),
pen (female) slightly less;
distribution: on most slow-
flowing waters.

You will find mute swans wherever there are sizeable expanses of relatively still water. These majestic birds are at home on slow-flowing rivers, lakes, gravel pits, large ponds —even in heavily built up areas such as London, where they add a touch of serenity to stretches of the Thames. Their haunts are not, however, restricted to freshwater, since they also frequent harbours and estuaries and have been sighted as far offshore as the Solent on the south coast of England.

Striking appearance The mute swan is one of the most easily identifiable of British birds. The adult's plumage is pure white and the thickly feathered neck is long and curves in a graceful 'S' shape. The head is small with a

down-pointed, orange-red beak tipped with a black nail, and there is a black knob over the nostrils at the base of the bill; this knob is more pronounced on the male than the female —and most obvious in spring. The webbed feet are black. The young swan has a greyish plumage which begins to turn white during the first winter and becomes pure white by the third spring, while the beak takes two years to assume its bright orange-red colour.

The swan uses its long neck to feed on underwater plants; pondweed and semi-aquatic plants make up the bulk of its diet. It also eats algae and shore plants and occasionally will take worms, insects and fish; it swallows grit and fine gravel for

Taking to the air
An adult swan is too heavy to take off from a standing position. To gain momentum it runs along (on land or on the surface of the water) with its neck outstretched and its wings thrashing violently. In a strong wind, it may have to run as far as 100m (300ft) before taking off. In flight the wing beats make a loud rhythmic noise. In order to land, the swan slows itself down by spreading out its wings and using its feet as brakes.

roughage. In deep water, where its neck is not long enough to reach certain food, it upends like a duck.

Its calls are quite out of character with its name and appearance, since it will snort, grunt and hiss threateningly when provoked; it also gives a shrill, trumpet-like call—particularly to the young who reply to the adult in a high-pitched tone.

Mating breeding and nesting Mute swans pair up, often in the autumn, when they are between two and four years old. Paired swans are not gregarious, preferring to nest isolated in their own aggressively defended territory. Non-breeding individuals and those of pre-breeding age may congregate together in areas where there is plenty of food and space. In spring, when their courtship and mating rituals reach a peak, you can see a pair of swans facing each other, swaying their heads sideways or dipping their heads in the water, extending their necks and bills vertically and even upending. Once the pair has been established, the cob (male) and pen (female) return annually to the same territory to breed. Swans seldom change their mate, unless for some reason they fail to breed.

The cob selects a nest site close to the water's edge and well away from other nests. Building the nest is mainly the pen's task, although the cob helps by gathering vegetation, often from previous nests, and passing it to her. The pair makes little attempt to camouflage the nest, which is a huge pile of reeds and sticks lined with a thin layer of down.

The pen lays her chalky, round-ended eggs every other day for up to 12 days; this occurs any time from April to July. The pen does most of the incubating, although the cob will take his turn and keep guard over the nest.

The eggs hatch after 36 days and the pen carries the broken shells to the water's edge. The young, which are born with their eyes open, are covered with soft, fur-like grey down; this is replaced by woolly feathers which change slowly to a drab brown colour. At five days they are independent enough to leave the nest during the daytime, although for up to a month they may continue to gather in the nest at night. Cygnets will walk long distances to water, marching along in single file behind their parents. If you are lucky, you may see one riding on its parent's back between the arched wings. They fly at four months and are usually driven away from the nest area in the following spring, when their plumage has changed from grey and when territories are redefined and their parents begin to prepare for the next brood. After leaving the nest the young join the summer flocks of non-breeders until they are ready to mate.

Less than half the swans in the British Isles are breeding stock. You can easily identify the non-breeders, which are immature birds or those which have yet to form a nesting pair, since they have pale pink bills and small nostril-lobes.

Moulting and migration If you come across a scattering of white feathers in July or

Below: This mute swan is in an aggressive mood as it protects its eggs (which lie in the large platform nest of reeds and sticks) from intruders. The wingspan can reach 2m (7ft).

August, do not assume this is the result of a fight; swans moult their flight feathers in the summer after nesting. When moulting they are vulnerable because they cannot fly.

Breeding adults, whose young cannot yet take to the air, moult near the nest site. Non-breeding birds gather in moulting flocks on quiet lakes or estuaries in June. The largest, of over 800 birds, is on Chesil Fleet in Dorset, while others of over 500 birds include those at Abberton Reservoir in Essex, the Ouse Washes, the Tweed estuary, and the Montrose Basin, Grampian. They moult 15 to 30 days after arrival, losing first their primary feathers and then the secondaries. They are able to fly again after a month, four weeks before the feathers have reached full length again. The return journey is made in September or October.

Swan ownership The mute swan is frequently described as a royal bird, and in fact all swans were once proclaimed to belong to the Crown, just as forests were reserved for royal deer hunting. Noblemen and city guilds were also given a right to own swans; a marking system of cuts and nicks on the swan's bill was used to identify the ownership of each bird.

Today the Dyers and the Vintners City of London livery companies are the last lawful owners (apart from the Queen). The Dyers use a single nick on the side of the bill, while the Vintners use two. The Queen's birds are not marked at all. This ownership of swans is taken as applying only to the Thames; elsewhere swan populations are unmarked and are treated as wild birds.

Swan upping Each July the colourful ceremony of swan upping—when the cygnets are rounded up and marked—takes place on the Thames. Three groups of swan uppers—one from each livery company and one led by the Queen's swanherd—row for four or five days up the Thames, from Maidenhead to Pangbourne, herding each brood of cygnets and marking them according to the marks found on their parents.

Threats to the swan The swan has no serious natural predators, although foxes and pike will sometimes take unguarded cygnets; and pike have been known to drown adult swans by holding their heads under water. The only real threat to the swan comes from man's increasing encroachment on its natural habitats. Overhead power cables are a major hazard to swans in flight and a number also die every year from oil pollution and mercury poisoning.

A common cause of death is lead poisoning. Swans often become entangled with fishing tackle and swallow the angler's lead weights. They also swallow lead shot and fishing hooks which have been discarded in the water at the end of the fishing season – a worrying result of modern disposable fishing hooks, lines and weights. The use of lead weights has been forbidden throughout English and Welsh waters.

An adult swan that is fortunate enough to escape all those threats to its life, however, can live up to 15 years and sometimes even longer; but on average a swan that succeeds in fledging will only survive for two to three years.

Above: The mute swan establishes a territory at the start of the breeding season and defends it aggressively, driving away intruders with a threat display or, as here, by chasing the other bird off.

Opposite page: There are few sights more majestic on a river than that of a swan sailing by, its long neck held in a proud curve.

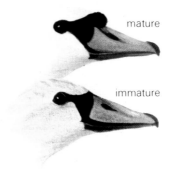

mature

immature

The mute swan's head is small compared to the rest of its body. The down-pointed, orange-red beak is tipped with a black nail. The characteristic black knob over the nostrils does not develop until the bird becomes fully mature.

ALL-YEAR TEAL AND SUMMER GARGANEY

Teal and garganey are closely related species of small dabbling ducks, with two big differences: teal are 50 to 100 times as common as the rare garganey, and while teal are seen all year round in Britain and Ireland, garganey are summer visitors.

Above: Two teal ducklings taking the sun on a mossy resting place. The nest itself is a shallow cup, usually quite close to the water, in which the female lays between 8 and 11 eggs. She incubates them for about 22 days, and the young fledge in a further four weeks.

Left: The male garganey's white head stripe contrasts with its reddish-brown head, neck and chest. The underparts and sides are pale grey.

The teal and the garganey are the two smallest ducks regularly occurring in Britain. Both are dabbling ducks, obtaining their food from the surface of the water or by up-ending; the males are brightly plumaged, and the females a dull camouflaged brown. They are each about 38cm (15in) long and fly with very rapid wingbeats and considerable agility, jumping straight from the water when disturbed.

Green-striped teal The male teal has a distinctively marked bright chestnut head, with a bold green curving stripe on each side surrounding the eye. This broad stripe is outlined in cream, and runs from the eye to the nape. The back and flanks are a delicately patterned grey, divided by a broad white stripe following the line of the closed wing. The chest is spotted brown, and the belly white. The tail is black but the area just in front is creamy-buff. The bird has a black bill and grey legs. The female teal is brown all over, except for a pale eyestripe.

Mixed populations Some 3000–4500 pairs of teal breed in Britain and Ireland. They occur almost throughout these islands, but are more common in Scotland and northern England than elsewhere. They favour rushy or boggy areas of moorland with numerous small pools, such as occur widely in the upland areas, and they also breed beside lowland lakes, rivers and streams, provided there is plenty of thick vegetation in which to conceal their nests.

These native breeding teal are probably largely resident, only moving small distances in the course of a year. However, teal breeding elsewhere in Europe are highly migratory, and many tens of thousands come to Britain for the winter. Peak counts in Britain have reached as high as 102,000, though normal numbers are probably about 90,000.

Wintering teal in Britain and Ireland favour shallow waters, both fresh and salt. Some of the largest concentrations occur on estuaries, for example the Mersey estuary in the north-west, the Essex marshes in the east, and in some of the south coast harbours. The Mersey estuary population is quite exceptional, and counts there have exceeded 35,000 birds.

Food requirements Teal feed mainly on small seeds and insects, and they obtain these by dabbling in fine mud and very shallow water, for example in estuaries and flooded fields. They walk or swim slowly forward, their bills submerged in the water and mud, filtering food items as they go.

Hard weather movements Extensive catching and ringing of teal in different parts of Europe have revealed that the birds respond quite markedly to varying weather conditions. Spells of unusually cold weather cause the most marked teal movements, driving them away from their normal wintering range, often pushing them south and west to areas beyond the extent of the frost and snow.

One example was the notorious hard winter of 1962–63, when the freezing weather lasted

Identification of teal and garganey

Teal (*Anas crecca*). Small dabbling duck. Resident, with large winter visiting population. 36cm (14in).

Garganey (*Anas querquedala*). Small dabbling duck. Summer visitor. Rare, with below 50 pairs. Virtually restricted to southern England. Length 38cm (15in).

reddish brown head and neck

white stripe from eye to nape

whitish underwing
dark leading edge
whitish eye-stripe

mottled throat and sides   pale belly ♀

Garganey

pale blue forewing

prominent scapulars

green stripe from eye to nape

grey back

chestnut head
white stripe
white wing bars

yellow patch

green and black speculum

green speculum
white wing bars

faint eye-stripe

Teal ♂

white stripe on underwing
whitish eye-stripe   faint eye-stripe

green and black speculum

dull green speculum

Garganey ♀   **Teal** ♀

Teal and garganey distribution

☐ Teal
■ Garganey

throughout January and February in Britain and Ireland. It is thought that the majority of teal wintering in Britain moved westwards into Ireland first, and then turned south for France. Even the west coast of France was not spared the grip of cold conditions, and so the teal had to move on even further, eventually reaching Spain and Portugal.

In all European countries there is an open season for shooting game birds and wildfowl –from around September to January or February. In several countries, including Britain, shooting is banned if there is a prolonged period of cold weather. This is done to reduce unnecessary mortality at a time when the birds may have become weak and confiding for lack of food.

However, the system has one major weakness: shooting may be stopped in the normal wintering range of the teal, for example in Britain and the Netherlands, but outside this area the need for the shooting ban is not appreciated. If the weather remains mild in France, shooting continues. Yet it is here that tens of thousands of teal are taking refuge –unfortunately, under exceptionally severe shooting pressure.

White-striped garganey The most obvious feature of the male garganey is the curved white stripe running from in front of and above the eye to the nape. It contrasts with the reddish-brown head, chest and neck, and pale grey sides and underparts. The female garganey is the usual camouflaged brown of most female ducks, but she has a noticeable whitish eye-stripe, a white spot at the base of the bill, and a white chin.

The garganey is the only species of duck that is a summer visitor to Britain, arriving in April and departing in September or October. It winters south of the Sahara, in West and East Africa. In Britain, the bird is more or less confined as a breeding species to the southern half of England, most pairs breeding in East Anglia and Kent. It is not common anywhere in the country, and has been declining in recent years. It was estimated that there were between 50 and 100 breeding pairs in the late 1960s, but since then the total has fallen below 50 pairs.

Above: Two garganey drakes. The garganey favours the same kinds of shallowly flooded fields as the teal, but is less often seen on the coast. It nests in thick vegetation quite close to water. Its food consists of a wide variety of seeds and aquatic insects.

Right: Teal frequent rivers, streams and shallow lakes. They also visit areas flooded by heavy rain. When reservoirs are being filled for the first time, large flocks of teal often move in and feed in the gradually rising waters as these flush insects and seeds from the vegetation.

Goosander distribution

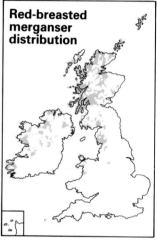

Red-breasted merganser distribution

SAWBILLS: GOOSANDER AND MERGANSER

The goosander and the red-breasted merganser–two quite similar fish-eating ducks–are often called sawbills. They get this name from the tooth-like serrations along the sides of their long, narrow bills. These savage-looking features enable them to take a firm grip on their fishy prey.

Above: Both the goosander and the red-breasted merganser breed in northern Britain, but only the latter breeds in Ireland.

Above left: A female goosander, followed by ten young: she is probably fostering several broods.

Below: The male goosander has a greenish-black head and back, with white underparts.

The goosander is a large bird, about 65cm (26in) long, with a wingspan of up to 95cm (38in). The male has a black head, neck and back, with a glossy green sheen. His head bears a distinctive crest of feathers, and his long bill and his legs are reddish orange. His underparts are pure white, contrasting strongly with the head and back. The rump, belly and tail are grey. In flight, even more white shows, as virtually the whole of the inner half of the wing is white, with just a few black streaks; the outer half is black. The female is grey above and white below, with a chestnut head and neck, and a slight shaggy crest. In flight the rear half of her inner wing is white, and the forewing grey.

Arrival of a species Goosanders breed in northern latitudes throughout the Northern

Hemisphere. Rather confusingly, they are known in North America as common mergansers. It has been estimated that there are about 75,000 goosanders in north-west Europe, with between 1250 and 1500 pairs breeding in the British Isles, mainly in Scotland. Yet the first definite breeding record of a goosander in Britain was as recent as 1871, when single pairs bred in Perthshire and in Argyll.

There had been rumours of breeding attempts earlier than that, in 1858 and again in 1862, but certainly the goosander was a rare winter visitor to Scotland before the 1870s. However, from about that time it became increasingly common, with winter influxes presumably leading to birds staying for the summer and breeding. By the turn of the century it had spread to most western Scottish counties, from Sutherland to Argyll, and was moving eastwards across the country.

During the first part of this century, breeding spread to the east coast counties of Aberdeen and Angus, and southwards to Renfrew, Selkirk and Dumfries. In 1986 it was estimated that there were almost 1000 pairs in Scotland.

Summering birds were present in Northumberland from the late 1920s onwards, but it was not until 1941 that the first breeding record for England occurred in that county. The first nesting in Cumberland was in 1950, and during the next 10-15 years the birds spread south into County Durham, Westmorland and Lancashire, and bred in North Wales from about 1968. Today there are about 150 pairs breeding in Northumberland, with at least as many pairs in other English and Welsh counties.

This increase and spread, welcome as it might be to birdwatchers, has been met with considerable opposition from fishermen, river bailiffs and others involved with trout and salmon fishing. It is claimed that goosanders cause great damage by taking young fish, and the birds are widely shot under licence, although there is no direct evidence that they are harming the fish populations.

Two fish-eating ducks

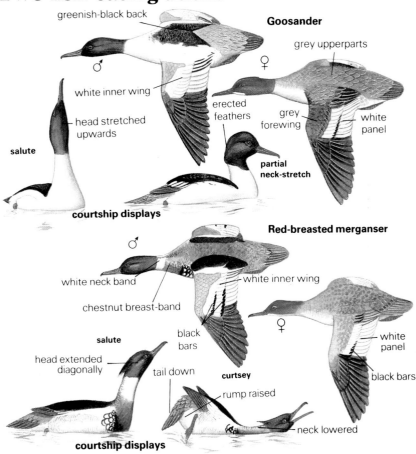

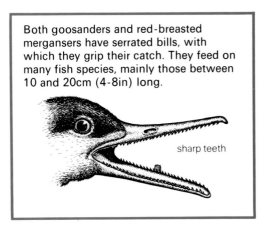

Goosander (*Mergus merganser*). Resident duck, nesting beside upland rivers and lakes. 65cm (26in).

Red-breasted merganser (*Mergus serrator*). Resident duck, mainly coastal. Up to 58cm (23in).

Below left: The red breast of the merganser is really bright chestnut, while the female goosander (below) has a chestnut head.

Both goosanders and red-breasted mergansers have serrated bills, with which they grip their catch. They feed on many fish species, mainly those between 10 and 20cm (4-8in) long.

Lakes and upland rivers Goosanders generally nest in holes in trees or banks, or among rocks, though they readily breed in nest-boxes and have used holes in buildings, from the cellar to the roof-space. The nest is usually within a few metres of water, though it can be up to a kilometre away. The female excavates a slight hollow within the selected hole, and lines it with whatever pieces of material or debris are within reach. She lays 8-11 eggs, and incubates them for about 30 days, the male taking no part in this, nor in rearing the young. The latter takes about two months.

Goosanders nest beside freshwater lakes, and along the upper reaches of fast-flowing rivers and streams. In winter they descend to the broader lower reaches, including estuaries, and also appear on reservoirs and gravel pits in small numbers. The principal food is fish, which they catch by diving, though they may first locate their prey by swimming on the surface with just the head and neck submerged. Underwater, they use only their feet to propel themselves, keeping their wings closed tightly. Dives can last for as much as two minutes, but half a minute is more usual, and they probably do not dive deeper than about 4m (2 fathoms).

The red-breasted merganser This duck is between 50 and 58cm (20-23in) in length, with a wingspan of up to 85cm (34in). The male has a black head, shot with green, ending in a ragged crest. His neck is white, contrasting with a broad chestnut breast-band. His upperparts are mainly black (with a green gloss) and grey, with grey flanks, though a broad white band shows on the closed wings. The underparts are white. The bill and legs are red. The inner half of the wing is white, but crossed by two black bars, and with a dark leading edge. The outer wing is dark.

The female is very similar to the female goosander, but has a larger shaggy crest and a duller chestnut head, and no sharp dividing line between the colour of her neck and breast. In flight, her wing pattern is more obviously barred black and white.

Expanding population The red-breasted merganser has a very similar distribution to that of the goosander, occurring all round the

Above: A male red-breasted merganser with his shaggy crest. Unfortunately, both goosanders and red-breasted mergansers are persecuted on Scottish rivers because they are believed to eat too many young trout and salmon. There is, however, little direct evidence of just how much damage the birds may be doing to fisheries. Despite extensive shooting, and destruction of nests, the birds continue to flourish.

Below: Red-breasted mergansers on the sea. They are almost purely fish-eaters, obtaining their prey by diving. Unlike the goosander, they use their wings to propel themselves, being one of the few species of birds that 'fly' under water.

Northern Hemisphere, but extending further north, well into the Arctic. There are thought to be about 75,000 birds in north-west Europe, of which about 2000 pairs breed in Britain.

The red-breasted merganser has probably always been a British breeding species, but at the same time that the goosander was colonizing the country, it underwent a marked expansion, in both range and numbers, which is probably still continuing. It now breeds throughout the northern and western parts of Scotland, extending south through south-west Scotland into north-west England, where it first bred in 1950. It is now quite widespread in Cumbria, particularly in the Lake District, and has bred in Wales since about 1953.

A coast dweller The red-breasted merganser is mainly a marine species, breeding around the coasts of Scotland, in sheltered bays and inlets as well as in estuaries. The birds also enter the lower parts of rivers, remaining separate from the goosanders, which breed in the upper reaches. During the winter they are found mainly on the sea, though sometimes occurring on inland freshwater lakes. Some large flocks are seen in certain estuaries, for example the Beauly Firth near Inverness, where over 3000 have been counted.

The nest is nearly always on the ground, concealed in thick vegetation, or among tree roots or in hollows and crevices on banks and low cliffs. Small islands are favoured for nesting, presumably for added safety from predators. The nest is rarely more than a few metres from the water. The female scrapes a shallow depression and lines it with any available grass or leaves. In it she lays 8-10 eggs and incubates them for about a month. Neighbouring broods may amalgamate, some being cared for by a single foster mother.

GREY HERON: TOP OF ITS FOOD PYRAMID

The grey heron is the top bird predator of the freshwater food pyramid–a status which affords it a wide range of diet, but also makes it most vulnerable to the build-up of toxic chemicals in the water. Despite this and other threats, the number of heron pairs in England and Wales remains at a stable 5000–6000.

Above: A magnificent bird with a wide wingspan of 175-195cm (68-76in), the grey heron has a white head with a dagger-like beak, and a broad black streak running from the eye to the tip of the crest.

The large, stately and distinctive-looking grey heron is generally found beside or in fresh water. You may also spot it flying with its long neck tucked in against its body and its legs stretching far beyond the tip of its tail. The harsh, honking 'fraank' call of the heron is a familiar sound on the marshes, instantly recognisable to birdwatchers and the many anglers who come across herons while they themselves are fishing.

Colonies Herons breed in colonies, preferring huge nests in trees; but sometimes they nest on cliffs, or at ground level on islands in lakes and lochs, especially in Scotland. A very few colonies, including the one at Tring in Hertfordshire, have nests in reed-beds –

sites more commonly favoured by the purple heron on the continent. Most colonies contain 10-30 nests; only one in the latest census exceeded 100 pairs.

Herons at a colony refurbish their old nests in January, or even December, ready for the early breeding season. Most colonies have traditional 'club' areas close by, where the 'off-duty' herons stand around. These gatherings are called 'sieges' and provide the sites where the birds begin their courtship display.

The male chooses a site and then advertises himself and his site to the females. When one of the latter shows interest, both sexes display together; this leads to mutual preening and the start of nest-building. The male generally

Grey heron (*Ardea cinerea*), total length about 95cm (36in), body length about 43cm (16in). Distribution beside almost any freshwater margin, especially marshes. Resident.

157

collects sticks, twigs, leaves, grass and other nest material and the female arranges them – in most cases using the base of the original nest. The birds mate on the nest or in the immediate vicinity.

From February right through to May, and even June, the female lays her eggs, which both parents incubate; two broods are rare.

Migration Herons often fly long distances to find new sources of food during cold spells, and are good at locating springs, sewage farms and other open water. In severe weather they also forage on the shore, as the salt in the sea prevents the water from freezing.

Herons from Scandinavia arrive in Britain for the winter as they are not able to survive this season in the colder parts of Europe. A few British herons leave for France, Spain or Portugal, but the majority remain within 100 miles of their native heronry.

Predator and pest Herons forage for a mixed diet of fish, eels, frogs, small mammals (quite often voles) and sometimes large

drin, accumulate in animals from the bottom of the heron's food pyramid upwards, and concentrations of these poisons increase until they reach the heron, which is the top bird predator of this pyramid. Analysis shows that the build-up of chemicals in the heron's body makes its egg shells extremely thin – to the point where, in some cases, they collapse or are broken by the weight of the brooding parent birds.

Population The effects of toxic chemicals and the gun put severe pressure on the heron population. In 1954 protection was granted against shooting, though it was sometimes difficult to enforce, and numbers returned to previous levels (as they also do after a fall in heron numbers caused by a hard winter). The most recent census figure for England and Wales stands at 5800 pairs, with a further 3000 pairs in Scotland. This makes a maximum population at the start of the breeding season of 20,000 birds. All these figures, which have been kept since 1928, are possible only because herons breed in colonies.

Right: A heron lays 3-5 chalky, blue-green eggs (surprisingly no bigger than hen's eggs), which hatch 3½-4 weeks later. The nestlings fledge after a further 7-8 weeks.

Below: The heron flies to a nearby field to beat its prey to death or swallow it.

Below: The heron spots its prey and stands motionless, with neck outstretched and head near the water ready to strike.

Left: The heron stabs its victim (in this case an eel) with a very fast, accurate blow. Just before the strike the heron often re-adjusts the position of its head, to make sure the final stab tells.

Below: All herons, including the purple heron (*Ardea purpurea*), can keep the head still, while swivelling the eyes downwards.

insects such as grasshoppers and emerging dragonflies. In the past, water bailiffs persecuted herons for their fish-eating habits, and trout farm owners today find individual herons persistent and expensive pests. Herons also welcome the source of food provided by suburban fish-ponds – goldfish are particularly easy to see.

With such a varied and abundant source of food, it is perhaps not surprising that herons have long lifespans. The record is held by a German bird that survived for 24 years; the two oldest British herons lived for 18 years.

Mortality Life can be hard for herons. All sorts of hazards may cause death – choking on large fish, being shot, becoming entangled in barbed wire, being killed by foxes, and (8% of all ringed recoveries) flying into overhead wires. The British Trust for Ornithology's sample census of heronries, which is the longest running survey of bird numbers in Britain, also shows how devastating the effects of severe winters can be.

Toxic chemicals, such as DDT and diel-

Relatives The grey heron is the only one of its group that occurs regularly in Britain, although its close relative, the purple heron, visits occasionally during the summer. Our native heron has a soft grey plumage, while the purple heron has darker, richer colours which are easy to distinguish. At a distance the young grey heron with its darker plumage is sometimes confused with the purple. The latter, however, is still very rare in Britain.

Apart from the bittern, which is a distant relative, the only other regular herons in Britain are the dazzling-white little egret and cattle egret – and both are even rarer than the purple heron.

Historical birds Heron colonies are generally in traditional sites – the oldest one, recorded in 1293, is at Chilham in Kent. The long history of this heronry is no accident. In the Middle Ages, falconry was an aristocratic sport, and herons were the most prestigious quarry. Herons also featured regularly as a *piece de resistance* at medieval banquets where they were often called cranes. It was therefore well worth noting the presence of a heronry in the land deeds. Heronries are commemorated in the names of houses, farms and villages – Herne Hill in London implies the hill with a heronry on it, and there are records to confirm this.

The bird's name has only recently been established as 'grey heron' and it is still known by local names such as herne and heronsew.

Preening

Preening is feather maintenance; as birds rely on their feathers for flying, for keeping dry and for insulation, they must keep their plumage in top condition. Close examination of a bird preening, scratching and shuffling, shows that these are very exacting activities. Individual feathers are teased through the bird's bill to ensure the barbs are aligned properly. The uropygial gland, situated above the base of the tail, provides natural oils for grooming, but not for waterproofing as one might expect. Herons (and bitterns) are particularly likely to get slime on their feathers from the food they catch. So they have a special tract on the breast (right), and two more on their flanks, providing a crumbly, powdery material ('powder down') which they use in preening. It is as if they are able to carry a natural, constantly renewed powder puff around with them.

Some birds, including nightjars that prey at dusk on moths whose wing scales then soil the birds, use their central claws to preen such scale material from their feathers. During the breeding season especially, you may see two birds preening each other. This could be helpful for feather maintenance, but it has a much deeper significance. It reinforces the pair bond between male and female (very like petting in humans) and is an important part of heron courtship.

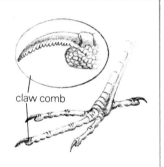

claw comb

Above: The heron uses its comb-like claws and the powder down located in its breast (below) to preen.

GREY WAGTAILS BY THE WATERSIDE

The grey wagtail is similar to the yellow wagtail, except for its blue-grey upperparts. This difference is 'clinched' by the habitat – the grey wagtail lives beside rocky streams, while the yellow wagtail prefers damp meadows and marshes.

The grey wagtail is somewhat confusingly named: although it is true that the upperparts of the bird are a bluish grey, the feature that most often catches the eye is the brilliant patch of lemon yellow on the under-tail coverts and on the belly. In summer this yellow colouring often extends up on to the breast. It is at all times conspicuous, even in birds seen flying overhead. Thus it would seem sensible to call this the 'yellow wagtail'. However, the true yellow wagtail has yellow upperparts as well as underparts; besides its difference in colour, the yellow wagtail is distinguished from the grey by the habitat in which it is seen, for it is a bird of damp meadows and marshes, only occasionally being seen on arable farmland, while the typical habitat of the grey wagtail is beside fast-flowing water.

Changing with the seasons Unlike yellow wagtails, which are migrants visiting us in summer from Africa, most grey wagtails remain in Britain or Ireland throughout the year. Their plumages, however, change with the seasons. In summer, the male grey wagtail

Above: The grey wagtail's nest is built by the female (shown here). She performs the bulk of the work of incubation, which takes 13 to 14 days. However, both sexes feed the young. In many lowland areas, two broods are raised, and the season usually starts early. Most pairs produce their first eggs in the second half of April. Natural nest sites include cavities between exposed roots, wall crevices (as shown here), and holes in the banks of streams.

Right: A grey wagtail nest with 5 eggs. Apart from this rather unusual nest site, other man-made sites used by grey wagtails include mills, bridges and weirs. Indeed, a good way to find the birds is to seek out old mills, which are favoured haunts.

has a conspicuous white stripe above and below the eye, and a large black 'bib', while the female's throat is whitish. In winter, the sexes are difficult to separate, both having buffish-white throats and pale buff breasts, with the yellow restricted to the vivid patch below the base of the tail. Young birds are grey-green above and dirty white below, with some dark marks round the throat.

Beside rivers and streams The map of the grey wagtail's breeding distribution shows a uniform coverage of Ireland, much of Scotland, Wales and western England. The only 'blank areas' are in East Anglia, the easternmost Midlands and Lincolnshire. In the western part of its range, the grey wagtail is typically a bird of shallow, fast-flowing rivers and streams, preferring to feed along the waterside or where the water is broken up by pebble banks.

Metallic call Often the first indication of the presence of grey wagtails, if they are not seen flying in an undulating path low over the water, is their frequently uttered call. This is considerably higher-pitched and more metallic than that of the pied wagtail. The grey wagtail's 'tzi-tzi', although formed of two distinct syllables, lacks the harsh emphasis of the pied wagtail's 'chiz-zick' call.

In mild winters, the song starts soon after Christmas. It is often said that the bird sings infrequently and even stops singing early in spring, but this may be because the sound is difficult to hear well against the background noise of fast-flowing, tumbling water. The song, in fact, sounds like a squeaky version of the descending trill of a blue tit's song, or perhaps a treecreeper's song without the terminal flourish. This is so different from the song of other wagtails that if the singing bird cannot be seen, the sound can cause moments of uncertainty even for the experienced ear.

Varied diet The grey wagtail's food consists of small animals – insects for example – that the bird gathers as it dashes nimbly among the pebbles and boulders, or runs across floating water plants. Sometimes a grey wagtail seems to tiptoe along the brink of a weir, risking being swept away by the rushing water. At other times (particularly

Grey wagtails in summer

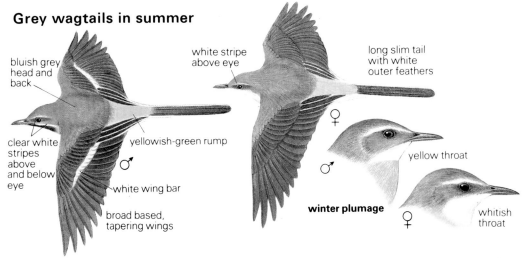

bluish grey head and back

clear white stripes above and below eye

white wing bar

broad based, tapering wings

yellowish-green rump

♂

white stripe above eye

long slim tail with white outer feathers

♀

♂

yellow throat

winter plumage

♀

whitish throat

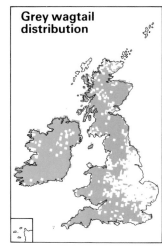

Grey wagtail distribution

when columns of midges are 'dancing' in their mating flight) grey wagtails fly out over the water and hover, as flycatchers do, to snap insects from the air. Besides aquatic and terrestrial insects, the birds take a wide variety of small animals, including worms, leeches, flatworms and small freshwater shrimps.

Surviving winter Because their habitat is along the waterside, where a cold winter freeze either kills or covers in ice any animals that the birds need for food, grey wagtails tend to suffer considerable losses in winters with extended and severely cold spells. Upland birds can 'migrate' to lower altitudes in winter without difficulty, but once the lowland streams and lakes freeze, only those birds that can reach a suitable stretch of unfrozen coast can survive. This may help to explain the distribution pattern, with more birds in the west, for the slow-moving streams in the generally colder east freeze quicker, and for longer, than the faster running streams of the west. This makes it rather difficult to

estimate the grey wagtail population: numbers are subject to sweeping variations caused by cold-winter mortality every few years. In 1967, following two severe winters, the population was estimated to be below 10,000 pairs, which is probably the lowest level to which it ever falls. There are now estimated to be about 25,000-50,000 pairs, thanks to milder winters.

Periods of bad weather provoke long-distance winter flights (better called cold weather movements), and ringed birds have been recovered in France, Spain and Portugal. Most flights are shorter, however. Many birds leave their home stretch of water during the period of moult in late summer, and at this time there are records of ringed birds being recovered up to 160km (100 miles) away from their nesting territories. For much of the year, most grey wagtail pairs stay in their territories and defend them: often both sexes defend their territory from a neighbouring pair, with vigorous posturing and mock combat.

Above: Numbers of grey wagtails are considerably lower in eastern England, where the streams are generally more sluggish and offer few suitable habitats.

Above left: The grey wagtail is distinguished by its combination of blue-grey upperparts and yellow underparts. The extent of the yellow is greater in summer than in winter, and in older birds compared to younger ones.

Grey wagtail (*Motacilla cinerea*). Resident bird living beside fast-flowing streams. Moves south in autumn. Length 18cm (7in).

Right: A male grey wagtail lives up to its name—the wagging of the tail neatly recorded by a blurred stripe on the photograph. There has been a considerable amount of debate concerning the reasons for the tail wagging in the wagtail family, with no certain conclusion. However, one theory that is at least plausible is that against the sparkling, ever-changing background of broken water, the continuously moving, white-edged tail helps to break up the silhouette of the bird and give a good camouflage effect.

BATTLING COOTS AND MOORHENS

Coots and moorhens are waterside birds of rivers and lakes but, unlike the ducks, they spend much of their time feeding on land, running about on their disproportionately long feet. In the breeding season both species become territorially aggressive.

Below: A moorhen among the grasses at the edge of a stream. The bright red beak 'shield' and the white area under the tail distinguish it from the coot. Another characteristic is its habit of jerking its head back and forth as it swims.

The coot and the moorhen, close relatives belonging to the rail family, are common and widely distributed throughout the British Isles. The coot is all black, with a white 'shield' above its white bill, and a faint whitish wing bar (as it does not fly very much, however, this last feature is rarely seen). Its legs are greenish in colour. The moorhen is brown-black in colour, tending to grey on the sides and underparts. Its shield, smaller than that of the coot, is bright red, as is its bill. There is a whitish bar down the flanks, and the area under the tail is also pure white. Its legs are green.

Both coots and moorhens can become totally accustomed to man. Moorhens, especially, breed beside the smallest farm or village pond, and also appear throughout the year on lakes in city and town centre parks. At a distance, both species can be confused with ducks. The moorhen, however, is smaller than most ducks, raises its tail well out of the water, and jerks its head backwards and forwards as it swims. The coot has a rounded, dumpy outline – very different from the straight-backed appearance of most ducks. Once you see the short, pointed bill you can easily distinguish both birds from the broad-billed ducks.

Territory, nests and chicks During the breeding season coots and moorhens defend territories against others of their own species. The area of the territories varies greatly, but

Like ducks, coots swim well and dive for food. Instead of having duck-like webbed feet, however, the coot has broad lobes on each toe (see below) which help to propel the bird through the water. (The moorhen is less aquatic in habits than the coot and has no lobes on its long toes.)

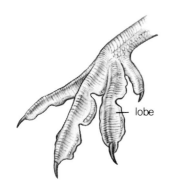

lobe

the coot usually defends a much larger one (up to an acre or more) than the moorhen. Some moorhen territories may be less than 100 square metres (123 sq yards) in size. The territories provide all that the pair needs in the way of food and nest sites.

The nests of coots and moorhens are built of dead leaves and the stems of rushes and other water plants. They are often placed in shallow water among concealing vegetation, sometimes anchored to the bottom, but often just floating on the water. Moorhens, more than coots, also build on dry land, occasionally a little way off the ground in thick bushes. A shallow cup at the top of the nest is lined with green leaves and finer material. Both species lay between six and nine pale buff eggs that are liberally spotted with brown.

Young coots and moorhens are able to walk and swim within a few hours of hatch-ing; moorhen chicks tend to stay close to the nest for the first few days at least, returning to it to be brooded by the parents at night or in cold, wet weather. Around the time of hatching the adult male coot builds one or two 'platforms' within the breeding territory. These are similar to the nest but flatter on top and are provided with one or more ramps leading up from the water. The young coots are brought to these platforms and brooded there rather than in the nest.

The young of both species are blackish in colour, with reddish heads. In the case of the coot this colouring comes from the tips of the down plumules, but the moorhen's head is bare—it is the skin that is red. This colour also extends to the bills and tiny shields of the young of both species. The bright colour in the middle of the reddish head shows as a central 'target' area for the parents who are feeding the chick. It is noticeable that as the young grow and begin to feed themselves, the reddish colour fades.

Above: Coots just taking off for flight after a run-up. The widely lobed toes assist the bird in its laborious take-off from the water surface. To gain the necessary speed and impetus the coot has to patter over the water for several metres. It always looks rather laboured in flight.

Coot (*Fulica atra*); 38cm (15in) from beak to tail; distribution widespread, on rivers, lakes. Resident.

Moorhen (*Gallinula chloropus*); 33cm (13in) from beak to tail; distribution widespread, on rivers, streams, ponds, lakes, marshes. Resident.

Moorhen

red shield

yellow tip

white flash on flanks

white undertail area

red head

chick (second brood)

juvenile (first brood)

white shield

floating nest

Coot

red head

chick

Threat and aggression

prominent white
undertail feathers

rearing
posture

outspread
wings

fluffed plumage

threat display

outstretched neck

fighting

Above: The outstretched neck, raised wings, fluffed plumage and conspicuous presentation of the red shield (left) are typical moorhen display postures. If neither bird gives way after display, fighting begins. The birds come together breast to breast, rearing up on their tails to strike at each other with their feet (right).

Below: The threat display of the coot.

In defending their territories, both species become aggressive, displaying vigorously at intruding birds and often fighting with them. The white frontal shield of the coot, and the red one of the moorhen, are the principal 'signals' used to indicate aggression. The bird holds its head low over the water, with neck outstretched, then fluffs up its plumage and slightly raises its wings. As well as making the bird look larger, and therefore more threatening, the shield is effectively presented to the opponent against a background of black feathers.

Two moorhens in a dispute, perhaps at the boundary of their territories, not only fluff themselves up to show off their

shields, but also cock up their tails and swim round so that their brilliant white undertail feathers show to maximum advantage. If neither bird gives way after this, they rear up on their tails, coming together breast to breast and striking at each other with their feet. Sometimes the long claws interlock and the two birds tumble over and over, even rolling into the water (if the encounter was started on land), and generally making a considerable commotion. It is not uncommon for the mates of the fighting birds to join in so that all four moorhens form a flapping, grappling mêlée. The fights can end in one of two ways. Either one bird, or pair, gives way and is chased off by the victors, or the combattants mutually decide that enough is enough, and they break off and retreat slowly from each other, hunched and ready to resume battle at the slightest sign of further aggression.

The coot indulges in similar fighting, preceded by showing off its white frontal shield. It also raises its tail and swings round like the moorhen but, as the underside of its tail is black, the significance of this is not so obvious. Fights between four birds—two pairs—occur as with the moorhen. All four may be involved in one scrap, but not uncommonly the two male birds and the two females have separate fights alongside each other.

Aggression and fighting is very important as each pair of birds tries to establish a territory large enough for its breeding and feeding needs. Actual damage during the fights is rare: as in most close encounters between birds, the winner, if there is one, is the stronger or older bird, and the loser gives up before it gets injured. Researchers have found some evidence to suggest that the same pair of birds return in spring to hold and defend the territory they had the previous year. This certainly happens with some moorhens, but it has not yet been conclusively proved that the same thing happens with coots.

Above: Coots walking on ice. They rarely breed on waters of less than an acre in extent, but in winter appear in flocks on many town park lakes, becoming completely tame and joining the swans and ducks in taking food from people. The rounded, dumpy silhouette of the coot, clearly visible here, distinguishes it from most ducks, which have straight backs.

Below: Coot chicks hatching. The red colouring on the head seems to reinforce the strident calls of the chicks as they beg for food from their parents. When the chicks beg they raise their heads towards the parents and sway them slightly from side to side.

Family affairs The coot normally rears just one brood of young in the year, but the moorhen often manages to raise two, and sometimes even three. In nearly all other species of birds that raise a second brood, the first brood youngsters move, or are driven, away from the nesting area. In the case of the moorhen, however, the young of the first brood help to rear the next set of chicks. Occasionally they may help to incubate the eggs, but more usually they assist in the feeding of the chicks after hatching.

It is obviously advantageous for the parents to have help in finding food and giving it to their chicks, but it does also mean that the territory has to be large enough for all the birds—parents, first brood young and second brood young—to be able to live and feed within it.

Seasonal movements Moorhens in the British Isles and north-west Europe are almost entirely sedentary. Movements of ringed individuals have rarely been recorded over more than about 10 to 15 miles. Further east, however, where the winters become very icy, moorhens are forced to move away south and west to warmer areas, but few if any come as far as Britain. Many moorhens stay on their breeding territories throughout the year.

British coots, too, are with us all the year round, but are joined by fairly large numbers coming from the Continent to spend winter here. Ringing has shown that these birds come from as far afield as western Russia, as well as from the countries round the Baltic and nearer Britain. For a bird that always looks very laboured when in flight, coots travel surprisingly long distances on migration. Unlike the moorhen, coots spend the winter in large flocks, with few birds staying on their territories. On some of our reservoirs and gravel pits, flocks of up to 5000 may build up.

Diving for food Moorhens are omnivorous birds, eating a wide variety of plant leaves, stems and seeds, as well as molluscs, worms, the larvae of many different freshwater invertebrates, and whatever the generous public feeds to them. They feed on land as much as on the water. When swimming they often dip their heads under water, and even upend, but only very occasionally dive.

Coots, on the other hand, dive quite competently, first pressing the air out of their plumage, and making a little upward and forward leap to give themselves the necessary impulsion to submerge. Normally they go down only about 90 or 120cm (3-4ft), but greater depths, of 4.5m (15ft) or more, have been recorded. Coots are omnivorous, like moorhens, but tend to eat mostly plants. They graze short vegetation on land, and pluck aquatic plants growing in and under the water.

SAND MARTINS: BUSY TUNNELLERS

Sand martins are summer visitors related to house martins and swallows. You almost always find them near sandy cliffs, soft river banks or sand quarries, for here they dig their long tunnels, often in colonies numbering hundreds of birds.

Above: The young birds move to the outer end of the nest tunnel, where they await the return of their foraging parents. The large hole seen below the nest tunnel is not the work of sand martins, but probably results from erosion. It could be occupied by larger birds such as jackdaws.

Sand martin (*Riparia riparia*); summer visitor; nests in colonies in sandy cliffs; brown, with no blue plumage; 12cm (4¾in).

The sand martin, the house martin and the swallow are the only three members of the hirundine family of birds to be found regularly in the countryside of Britain and Ireland. Of these, the sand martin is the least well known; with its dull brown and white plumage, it appears at first sight to be the least interesting of the three. In fact it has a highly distinctive lifestyle, and its gregarious activities reward the birdwatcher with many memorable hours.

Distinctive brown plumage Sand martins lack the blue plumage of the house martin and swallow, and also lack the latter's long tail streamers. They are much smaller than their two relatives – about half the weight of a

swallow and two-thirds that of a house martin. They feed aerially and, when seen in flight silhouette, are very like the house martin, but their brown plumage is altogether different and distinctive.

Winter conditions in Britain are, of course, impossible for hirundine birds, as all their flying insect food is missing. The sand martins are therefore with us only from late March or April through to September or early October. During the winter they are in West Africa, most of them probably in Senegal.

Unusual colonies The earliest birds to arrive are the adults of the previous year. They may be three or four weeks ahead of the young birds, and are certainly well ahead of the swallows and house martins. The first arrivals are generally found in small flocks feeding over water, where the insect life is likely to be concentrated. Soon, however, they start to visit their old colonies, for sand martins are colonial breeding birds, nesting in vertical sand banks in which they excavate long tunnels.

Colonies, in natural situations, are typically found in riverside banks where winter floods have washed away the bank in sandy ground, leaving a vertical face. Sandy cliffs at lake or sea side may also be used. However, in the south of England such natural sites are very unusual and by far the majority of colonies are in man-made sites – where sand or gravel is being excavated, or where other excavations

Sand martin lookalikes

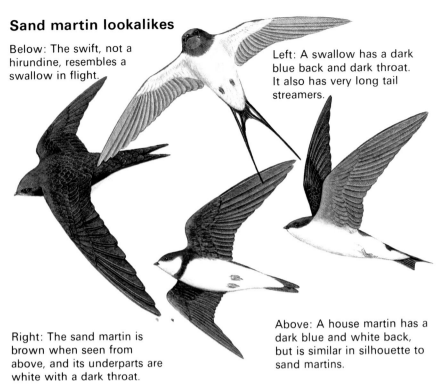

Below: The swift, not a hirundine, resembles a swallow in flight.

Left: A swallow has a dark blue back and dark throat. It also has very long tail streamers.

Right: The sand martin is brown when seen from above, and its underparts are white with a dark throat.

Above: A house martin has a dark blue and white back, but is similar in silhouette to sand martins.

Above: The clutch generally consists of five eggs (sometimes as few as three or as many as seven) and they are pure white. The eggs hatch after about two weeks and the young fledge some 18-20 days later. The very earliest nestlings, in a good year, fledge by the beginning of June, but in a well-populated colony, nesting is continuous until mid-August. The most successful pairs will by then have had three broods.

Right: This sand quarry has become the site of a moderate-sized colony of sand martins. Few successful colonies hold less than ten pairs of birds and, in years when the sand martin population is high, many colonies have hundreds of pairs.

have taken place. These may include railway or road cuttings, foundation diggings for buildings, silage pits or even heaps of graded sand. Fine soil in chalk or clay pits may be used, and there are also records of birds breeding in heaps of sawdust, in rotten brickwork and down the drainage pipes in hard masonry walls.

When breeding begins in a colony, birds are stimulated by the breeding activity of their fellow members, and so groups of sand martins are found all making their preparations at the same time. The early birds, working almost in synchronised timing, begin to excavate nest holes.

The nest chamber, situated at the end of the tunnel and slightly above the entrance for good drainage, is lined with fine grasses and feathers, which the birds generally catch in the air or, sometimes, take from the ground. One of the most exciting times in a colony is when the birds engage in feather games – one bird drops a feather and the others compete to catch it as it falls down the cliff face – until one manages to gather it, and then uses it in its nest lining.

Life of the young birds The young birds, when they fledge, have a much softer plumage than their parents, with brown or russet fringes to many of their feathers. The youngsters are lively and even disruptive, dozens of them taking part in extensive feather games. Soon they may be joined by young birds from colonies far away, who have started to move about the countryside. These flocks of moving birds roost either in established colonies where there are enough vacant holes, or they descend on a reedbed – something they frequently do in flocks of thousands.

As the summer progresses, the time soon comes when all the first brood young have started to explore their surroundings. It is

Above: In flight, sand martins reveal that their white underparts are marked with a band across the breast, distinguishing them from house martins, which have completely white underparts. To the dedicated birdwatcher, sand martins are as much a sign of spring as swallows are the harbingers of summer, for they begin to arrive from their wintering grounds during March. All through summer they can be seen making long, curving flights across open spaces such as meadows or stretches of open water, catching the insects that fly in myriads in these locations.

not long before they start their migration by gradually heading southwards. Within Britain, large-scale ringing has shown that the birds move a little east of south so that they can make a short sea crossing of the Channel. The roosts in Sussex, Kent and particularly in the Fens of East Anglia are an impressive sight, for they can hold tens or even hundreds of thousands of birds. Once across the Channel, the birds' progress has been charted along the Biscay coast of France around Nantes, and thence down to Spain and the Mediterranean – probably via the Ebro Valley. The few records in Africa come mostly from Morocco and then, in the winter, from Senegal.

Disaster and recovery The wintering area of the sand martin was subjected to a very severe drought during the 1960s. This area is known as the Sahel and lies on the southern fringe of the Sahara; the climate there has

always been rather dry, but a rainy season each year allowed trees and shrubs to thrive. In the summer of 1968 the rains failed altogether, causing a serious drought and altering the whole ecology of the area. Trees died, and the ground flora was badly reduced. This loss of foliage led to a fall in the amount of insect life, the food supply of the sand martin.

In Britain, the sand martin population had been at a high level through the early 1960s, but very few birds came back from Africa in the spring of 1969. It has been estimated that in the course of one winter, their numbers were reduced to less than a third of those of the previous year. A similar tragedy occurred in three other bird species: the whitethroat, the sedge warbler and the redstart, which all winter in the Sahel. Despite recent recoveries, Britain's sand martins are much less numerous than during the 1960s.

Communities on the cliffs Sand martin colonies house a community of bird and insect life. Many other species of birds breed in the holes – even little owls which may themselves catch and eat the sand martins. Tree sparrows, starlings and, in big old holes, jackdaws are probably the commonest guests of the sand martins. Many colonies are also regularly attacked by predators from outside – kestrels and sparrowhawks and sometimes even hobbies.

Besides predators, the community includes parasites, for the holes are a suitable habitat for certain species of fleas. Finally, crows, rooks and jackdaws occasionally join the community when they visit unoccupied holes to feed on the parasites.

The tunnel and its many tenants

The making of the tunnel The bird flutters and hovers at the face of the sand cliff and scratches with its feet, gradually starting the tunnel. As soon as the bird has a ledge on which to rest, it stands and kicks the sand backwards (far right). It may take two or three weeks to complete the tunnel.

Fellow inmates Sand martins carry numerous parasites. These include the minute feather-lice and the larger louse-flies. There are also feather mites, tiny creatures that feed on feather debris. Blood-sucking flies such as *Crataerina hirundinis* (inset) are also found among the plumage of sand martins. These insects, which are sometimes also found on swallows, swifts and house martins, have stunted wings and cannot fly. Like fleas, they are able to survive winter in a dormant state until the return of the birds, whose movements and bodily warmth reawaken the flies.

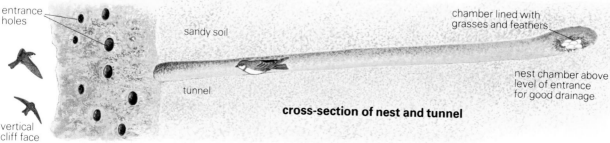

entrance holes

sandy soil

vertical cliff face

tunnel

chamber lined with grasses and feathers

nest chamber above level of entrance for good drainage

cross-section of nest and tunnel

BRILLIANT WATERSIDE KINGFISHER

The kingfisher is one of our most brilliantly coloured birds but its small size and rapid flight can make it difficult to spot. However, once seen it is seldom forgotten.

Below: Apart from its extremely distinctive colouring – no other British bird has brighter plumage – the kingfisher can easily be distinguished by its stumpy body, large head, short tail and long, dagger shaped beak. Years ago this bird was persecuted for the crime of poaching hatchery fish, but now it is a fully protected species under the Wildlife and Countryside Act of 1981.

In flight the kingfisher looks like a flash of bright blue light as it skims fast and low over the water. It is one of Britain's most beautiful birds, with upper parts of an iridescent cobalt blue – or emerald green depending on the angle at which the light catches them – and a very noticeable paler blue streak stretching from nape to tail. The underparts and cheeks are a warm chestnut colour, which is most obvious when the bird is perching, and there's a patch of white on the throat and sides of the neck. And as if all this colour were not enough for one bird, the legs are a bright sealing-wax red. Juveniles generally have a duller plumage, shorter bills with a white tip and dark legs.

169

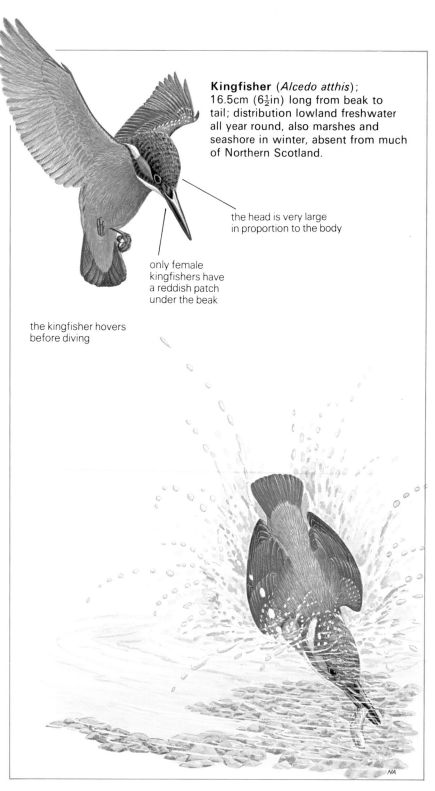

Kingfisher (*Alcedo atthis*);
16.5cm (6½in) long from beak to
tail; distribution lowland freshwater
all year round, also marshes and
seashore in winter, absent from much
of Northern Scotland.

the head is very large
in proportion to the body

only female
kingfishers have
a reddish patch
under the beak

the kingfisher hovers
before diving

Above: The kingfisher's characteristic method of catching prey is to perch on a convenient branch or tree stump until it sights a small fish; then it flies up and hovers directly above the spot where the prey is hiding before diving straight down into the water to catch it with its open beak. The kingfisher is fortunate in being seldom preyed on by other birds who avoid it because of the unpleasant taste of its flesh.

It would be surprising if such a colourful bird had escaped mention in myth and legend; the most attractive story comes from the ancient Greeks who said the bird – which they called 'halcyon' – bred in a floating nest at sea at the time of the winter solstice, calming high winds and stormy waves. Today 'halcyon' has come to mean any calm, peaceful and happy time.

Expert angler The place to spot the elusive kingfisher is near fresh, rather slow-flowing rivers, canals, lakes, ponds, streams, flooded gravel pits and even tiny streams in towns – almost wherever a ready supply of small fish such as minnows, bullheads and sticklebacks is to be had. These fish make up the bulk of the kingfisher's diet; but it will eat insects such as caddis fly larvae and dragonflies, as well as tadpoles, small molluscs and crustaceans such as crayfish.

As its name implies, the kingfisher is an expert fisher. It normally perches on a post or branch over the water, watching intently for fish; if a suitable perch is not available, it hovers over the water. The instant a fish is spotted the kingfisher dives headlong straight into the water and grabs its prey with its open, dagger-like beak; the complete action, from leaving the perch to returning to it, is over in a matter of seconds. Minnows may be immediately swallowed head first, but spiny fish such as sticklebacks are beaten against the kingfisher's perch until they are dead and their spines flat enough for the bird to swallow them comfortably. You might see the kingfisher juggling the fish in its beak to get it the right way up to swallow. On average, adults catch a fish once every two or three dives, while juveniles only catch one every eight or ten dives until they gain expertise. In learning the art, some young birds drown because they dive too often and their feathers become waterlogged.

Pairing and nest-building Throughout autumn and much of the winter individual kingfishers keep to their own territory, male and female using separate areas of water; but in January or February the pair bond is established or renewed. The birds chase each other, often at considerable heights, in a swooping, diving aerobatic display, or they perch on a branch bobbing and bowing to each other. The attentive male will even fetch fish to feed to his partner.

From mid-February to mid-April you may hear the infrequent song of the kingfisher – a rapid, high-pitched succession of varied whistles. The bird's normal and very distinctive call is a loud, shrill 'chee' or 'chikee'.

As the weather warms up the pair look for a nesting site, usually in an exposed bank of a stream or lake. Once a suitable location is found they excavate a tunnel, which they start by flying at the bank and driving in their strong bills; they then build a chamber at the end of it. The tunnel, anything from 15-100cm (6-40in) long depending on how hard the soil is, slopes gently upwards. If a suitable waterside bank is not to be had, the kingfisher will nest among the roots of a fallen tree or in a sandpit, sometimes as much as 300m (330yd) from the nearest water. If you spot a kingfisher flying through woodland, the chances are it will be fetching food to take to an inland nest.

You can recognize the nest by the circular entrance hole to the tunnel, positioned near the top of the bank. Later in the season, disgorged fish offal round the opening will provide evidence of a nest, as will the dark slime trickling from the hole – the excrement of the young which is directed into the tunnel from the nest chamber.

Raising a family Kingfishers lay eggs at any time between the end of March and early July. Pairs can rear two broods a season and sometimes even three. Male and female share in incubation, when the clutch – usually seven eggs – is complete. At first the eggs are a translucent pink but they turn a shiny white colour during the 19-21 day incubation period. Both adults diligently feed the young which are blind and naked when they hatch. The young are fed a diet of tiny fish at first. After about 10 days their eyes open and the first signs of feathers appear; after two weeks they can manage much larger fish. To encourage their parents to keep them well supplied with food, the young keep up a continuous trilling or purring sound; the hungriest stand at the front of the chamber calling for food while the more contented sleep at the rear. The accumulation of fish bones and regurgitated pellets, as well as the youngsters' excrement, make such a mess of the tunnel and chamber that the adults have to take a quick bath every time they leave the nest.

At the end of the breeding season the juveniles are chased out of the territory by their parents, who threaten them unmercifully and drive them out in noisy chases. Most go no more than six miles away, although some have been recorded as travelling up to 160 miles, mostly in the autumn, in search of fresh sources of food.

Hazards to survival Winter is an occupational hazard for all our resident birds and kingfishers are particularly vulnerable if the weather turns cold enough to freeze lakes and ponds and cut off the supply of fish. Many die of cold and starvation. Those lucky enough to find enough food to survive face an additional hazard – the alterations made by man to the watersides. Removal of trees used for perches, water pollution, the regrading of banks which destroys nesting sites and the levelling of streambeds which reduces the numbers of fish – all drastically affect the kingfisher's chances of survival. Fortunately, the kingfisher is still widespread over most of England, Wales and Ireland.

Above: Both male and female birds feed the young, who clamour incessantly for food; soon the nest chamber becomes littered with discarded fish bones.

Below: As part of the courtship ritual the male kingfisher offers his partner a fish. The female is the one with the reddish patch on the lower bill.

DIPPERS: HUNTERS IN THE RUSHING STREAM

Dippers nest beside fast-flowing streams and flit low over the water, alighting on favoured stones and boulders. They stand bobbing and bowing for a while, and then plunge into the rushing water to hunt aquatic creatures – they even walk beneath the surface, gripping the stream bed with their powerful claws.

Dipper (*Cinclus cinclus*) seen mainly beside fast, stony streams in upland areas; solitary dweller. Resident. Sexes look alike. Length 18cm (7in).

Below: The dipper is found in areas between 320 and 640m (1000-2000ft) above sea level; but in the north, in places where hills descend steeply to the sea, dippers nest close to the shore as well.

With its bold plumage and strong association with rivers and streams, the dipper is normally an easy bird to identify. Perhaps best likened to a 'thrush-sized wren', it has a jaunty, tail-cocked posture, and as it stands on some stone or boulder it frequently bobs up and down. Its legs are strong and look similar to those of a starling, while its beak is like that of a thrush in both colour and size. The upperparts are a rich, plain-chocolate brown, and the wings and tail look short in proportion to the distinctly plump body. By far the most conspicuous feature is the large, strikingly white 'bib' extending across the throat and breast, and down on to the belly.

In the case of English and most Scottish-breeding birds, this white bib has a broad chestnut margin which merges gradually into the near-black of the flanks and belly. Irish-breeding dippers (and also those in the Hebrides) have considerably less chestnut, and birds of the Continental race (which sometime stray to east and south-east England in spring and autumn) have almost no chestnut at all. The Continental race of dipper is in fact known as the black-bellied dipper.

Stony stream dwellers Just as characteristic as the plumage is the habitat of the dipper. Apart from stray migrants, which are sometimes seen on the coast or on slow-moving

Dipper distribution

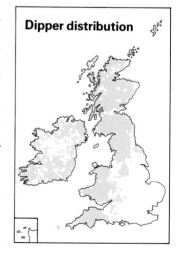

The British and Irish dippers

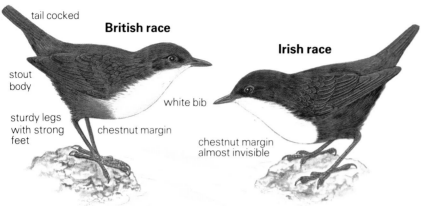

tail cocked

British race

stout body

Irish race

white bib

sturdy legs with strong feet

chestnut margin

chestnut margin almost invisible

often in mid-stream, and the dipper stands on them, bobbing up and down continuously. Against the background of broken white water, the clear white bib offers surprisingly good camouflage. As the dipper bobs and dips its tail, it also blinks, using its so-called 'third eye lid' or nictitating membrane, which travels fore-and-aft, rather than up and down, to wipe the eye surface clean. In the dipper, the nictitating membrane is conspicuously white. Under water, this membrane is thought to protect the dipper's eye from gritty particles; when submerged it becomes transparent and does not hamper the bird's underwater vision.

Underwater hunter From the boulders in

streams in the lowlands, dippers are typically birds of fast-flowing streams and the shallow rocky shores of lochs and tarns in hilly or mountainous country. In Britain, only those dippers that inhabit the south-eastern edge of the breeding range (roughly along a line from Swanage in Dorset to Flamborough Head in Humberside) frequent slower-moving lowland rivers or streams. These birds usually favour stretches near weirs or sluices, where the water flow is broken.

Most dippers are sedentary birds, establishing a territory on a suitable stream or river and rarely straying far from it. Those in the far north, or at higher altitudes, do descend to lower waters during the winter, but in general, severe winters (especially those with a sudden, rapid-freezing onset) are fatal for many dippers.

Dippers are usually to be seen singly or in pairs (except when they have newly fledged young), and most pairs maintain their territories through the winter months. They are reluctant to cross the 'border' with adjacent territories, and double back when reaching this invisible line. Usually their flight is low and fast, with frequent, penetrating 'zit . . . zit . . . zit' calls, generally over their home reach of water; occasionally, if alarmed, they fly overland.

Not only do they have a 'home reach' of water, but also a number of favoured boulders within this, from which they hunt. These are

Above: The dipper's nest is dome-shaped and built of grasses and mosses. Many dippers nest behind waterfalls, and nowadays in man-made sites such as mills, weirs or under bridges.

Below: In lowland territories, two broods a year are usual, the first eggs being laid in early March. In the Scottish Highlands, laying is up to a fortnight later, and often only one brood is raised.

its stream, the dipper hunts, often beneath the surface but occasionally just 'up to its waist' in the shallows. To submerge, the dipper may jump straight into deep water, or gradually walk into the stream until lost from sight. Once beneath the surface, it hunts on foot, and this is where the disproportionately stout legs and powerful, sharp-clawed feet serve a vital purpose in holding the bird's buoyant body down on the stream bed. Under water, the wings are used like the fins of a fish, both for propulsion and to maintain balance in strong currents—sometimes dippers swim for considerable distances.

Most small underwater animals form part of the dipper's diet at some stage, but favourite items include caddisfly larvae (in their 'portable homes' of tiny fragments of shell, stone and vegetation) and water-beetle larvae; worms and small shellfish; freshwater shrimps; the occasional small fish; and, in season, tadpoles.

The breeding season This begins early in the year. The tunefully sweet, if slightly disjointed, warbling song can be heard in the territories from October onwards through the winter, until towards the end of the nesting period in the subsequent June. Both sexes indulge in song with equal enthusiasm: yet another colourful feature of this attractive waterside bird.

Mammals of river and stream

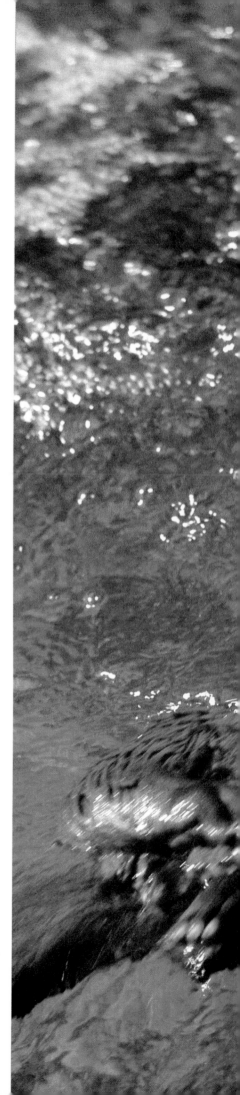

There are only three truly aquatic freshwater mammals in Britain – the otter, the water shrew and the water vole. The rest are either marine, like the seals, extinct, like the beaver, or introduced, like the mink.

Several other mammals can certainly be seen along rivers and canals, but they rarely take to the water by choice. Moles, for instance, frequently feed near rivers where the damp soil is a prime source of succulent worms, and foxes are drawn seasonally to canals and river margins to feed on the rich pickings of duck, coot and moorhen chicks, while the aggressive weasel has been observed swimming in pursuit of fleeing water voles. Apart from land mammals, bats patrol the air over large rivers and canals to take advantage of the insect life fluttering above the water.

The three truly aquatic British mammals contrast so markedly with one another in lifestyles that competition is out of the question. The diminutive water shrew is probably, for its size, the most ferocious animal in Britain. It is more active during the day than at night and tackles prey larger than itself with enormous relish and tenacity. Although it feeds most during the day, the water shrew is very difficult to see.

The same cannot be said for water voles, which are frequently seen swimming across rivers and canals, leaving characteristic V-shaped ripples in their wake. Even if they are not seen, however, they are frequently heard. A peaceful walk along the towpath of a canal is likely to be punctuated by plopping sounds as voles dive from the banks into the water.

Our premier aquatic mammal is unquestionably the otter, but unfortunately during the last 30 years it has declined in abundance drastically. The agility of this animal in water is incredible. Its smooth, bullet-like contours and the apparent bonelessness of its body allow it to execute underwater acrobatics which defy description. Despite this, the otter is a lazy feeder and prefers to catch eels and sluggish coarse fish rather than fast-moving salmonids. Its gymnastic underwater skills appear to be reserved for playful exploits – a clear reflection of the fun-loving nature of this endearing animal.

Left: The water shrew — the smallest of our aquatic mammals — is also the most ferocious. To keep alive it must eat just over its own body weight every 24 hours. Its life is therefore short but frantically active as it is almost unceasingly on the move in search of prey.

This checklist is a guide to the mammals you will find in or near rivers and streams. Although you will not see them all in the same place, you should be able to spot many of them throughout the changing seasons. The species listed in **bold type** *are described in detail.*

WATER DWELLERS
Mink
Otter
Pipistrelle bat
Water shrew
Water vole

VISITORS
Daubenton bat
Fox
Mole
Polecat
Rat
Weasel

Left: The otter is a playful mammal and tossing pebbles about is a popular pastime. This otter is holding a pebble between her dextrous forepaws like a small football, and will toss it into the air and retrieve it under water.

175

ELUSIVE OTTER OF THE WATERSIDE

Otters are fascinating animals to watch as they have many playful and endearing habits – sliding down muddy banks and twirling about in the water, for example. But because of their elusive nature and the fact that populations are dwindling, they are not easy to find.

Below: Otters feed mainly on fish and for centuries gamekeepers have waged war on them under the impression that otters damaged fish stocks. Otters certainly do take trout, and the occasional salmon, but they are very opportunist feeders and will also take frogs, tadpoles, and even water birds such as moorhens which they catch by coming up underneath birds as they swim, and pulling them under the water. Eels are also a common source of food.

The otter looks like an overgrown 'water weasel': and that's exactly what it is. Like its smaller relatives including the stoat and weasel, it has a long body, long tail and short legs. But it is the only native one whose search for food takes it into the water, and it is this amphibian existence which affects every aspect of its way of life.

Strong swimmer The otter's body is suitably shaped to move fast in water as well as on land. At one end its small head merges with its thick muscular trunk, while at the other end its powerful tail – or rudder – tapers from a broad base to a point. Its five-nailed toes are webbed like a duck's and transform an ineffectual dog-paddle into a powerful swim-

ming stroke. Altogether the otter is beautifully streamlined; it swims on the surface by paddling with its feet, and when swimming fast it flexes its supple body and tail. Sometimes when it is playing it may use a 'porpoising' action through the water.

Furry protection As the otter is frequently going in and out of water it needs a coat that can act as both mackintosh and warm blanket: this is exactly what its rich, thick fur provides. The coat is made up of two layers: the visible one is long and coarse, while the under-fur is fine, glossy and so thick that it is almost impossible to part. When the otter submerges, the under-fur traps a layer of air bubbles, which insulates the animal by preventing water getting in. This layer also creates the characteristic silver colour that otters have when they are swimming under water – and the trail of bubbles which marks their progress.

When the otter climbs out onto a beach or bank, water cascades off its body, and the long guard hairs form bunches, giving the coat a distinctly spiky appearance. To dry off it has a good shake, and often also rolls on the grass. The otter spends a great deal of time grooming and generally caring for its coat.

How does the otter stay under water in order to catch its prey? Just before it dives, it takes a very deep breath (sometimes a gasp is clearly audible). Because its lungs are large,

that single breath will keep the otter going for three or four minutes, giving the creature time to swim up to a quarter of a mile, catch a fish or escape most dangers.

The otter has other features adapted to underwater swimming. Its ears hardly project beyond its thick fur so as not to spoil the stream-lining of its body. When it is submerged, both its ears and crescent-shaped nostrils are closed by special valves. The otter can hear little or nothing underwater, but relies on its sight which is especially adapted for this purpose. Round the eyes are special muscles, rather like those of a cormorant, which apparently adjust the focus to compensate for the visual distortion caused by the water. The otter therefore seems to see even better in the water than out of it, and it certainly normally hunts by sight.

Sometimes when the otter is in choppy or murky water, even eyesight may be useless. The otter then relies on its whiskers. It has bunches of these on its cheeks, throat and eyebrows. These whiskers are so sensitive to vibration that an otter can chase fleeing fish in dark water.

Like other amphibious creatures such as the frog, the otter has rather bulging eyes positioned near the top of its flat head. This gives it a good view in front and above, but apparently not below, since it seems unable to catch fish swimming underneath it. Instead the otter usually comes up on its prey – a good tactic where fish are concerned, since they too can scarcely see what is going on below them. When hunting for eels (a favourite food) the otter will turn over stones at the bottom of the river with its paws.

Smelly messages Contact with other otters is maintained chiefly by scent messages in the form of a special anal jelly, produced by a pair of anal glands under the tail, and droppings (spraints). The full significance of these messages is not known, but research has shown that the chemical character of the jelly is as individual as a signature and that otters can distinguish between deposits left by different otters. The jelly may also be used by the dog otter to tell whether a particular bitch is 'on heat'.

The spraints are much easier to find than the special jelly. They are deposited in places where other otters are most likely to find them – on ledges, under bridges or on rocks in mid-stream. Good sites for depositing spraints will be used year after year. The spraints are dark in colour and have a very distinctive, not unpleasant, musky smell: once smelt never forgotten. They can be any length up to 10cm (3½in) and any consistency depending on what the otter has been eating. Examination of their contents gives a good idea of the otter's diet as the hard parts of its prey, such as fish bones, pass through the gut without much change.

Mating Otters usually move singly and mainly at night, with exclusive rights on an

average 15-mile stretch of river. They have more than one resting place (holt) between which they move in an unpredictable way. Although by day otters lie up on their own, a bitch on heat immediately becomes an object of interest to the dogs (males) in the neighbourhood. She will usually only mate with one otter but sometimes several may pursue her and fight fiercely for her favours. Amid much squealing the rival males snap at each other, particularly in the region of the genitals; broken penis bones are quite a common casualty among dog otters.

Even for the successful male otter his difficulties are not over. He must rendezvous with his bitch in the dark; often he does this by engaging her in a whistling duet. First one whistles, then the other; only when they have got a bearing on each other's position do they approach slowly. Otters mate whatever the time of year – usually in the water. Since pairing may last 15 minutes, in winter this is a considerable tribute to the effectiveness of the layer of air insulating the otter's body.

Breeding holt The couple usually stay together for a few weeks, but before the cubs are born the bitch drives the dog away. Two months after mating, the bitch gives birth. However, very little is known about where she gives birth or how she looks after her cubs in the wild.

Otters with riverside territories may well give birth in places similar to their 'lying-up'

Above: The otter is an alert and inquisitive animal – as the stance of this individual clearly shows. The long, sensitive whiskers assist it to catch fish and find its way in murky water.

Below: This characteristic pose of the otter is known as tripoding; balance is kept by the hind feet and the base of the outstretched, muscular tail. From this elevated position the otter can keep a good look-out.

holts, for example in old burrows, natural hollows in the riverbank, under old tree roots or hollows under boulders. The holt has to be situated well above the water level of winter floods and well hidden. Another possibility is that the cubs may be born in holts away from the water's edge so that there is no danger of them drowning before they are old enough to swim. The mother would then bring them down to the river as soon as they were old enough to be moved. Coastal otters probably make use of caves, while in fenland areas such as East Anglia surface nests made out of reeds have been discovered.

Vulnerable cubs The young – between one and five in a litter – are born blind, toothless and are about the size of a rat. In this helpless state they utter soft twittering cries which develop into chirping or chittering. The cubs are not weaned for four months, and grow relatively slowly. They stay with the mother for anything up to a year, during which time she rears them without the male's assistance, and defends them fiercely when necessary. In the past the female otter has won reluctant admiration from gamekeepers and huntsmen for the way that, even when wounded, she returns to her litter, an instinct which has sometimes caused her death and that of her cubs. Sometimes if her cubs are in danger, she may summon them with a sharp whistle. There are also reports of dogs and people being attacked because they have threatened the young or accidentally disturbed them.

The young leave the holt for the first time when they are about two months old; but they are nearly twice that age before they make their first excursion into the water, because their long silky fur is not yet waterproof. The cubs do not take to water naturally. The mother has to coax them and push them in, or even take them in her mouth and dunk them forcibly. Once in, however, the cubs are quick to adapt to their new watery home and are soon beginning to dive and 'rough and tumble' with one another.

Few animals enjoy a good game as much as otters seem to. They even play on their own as well as in company with other cubs or together as adults. In captivity and in such surroundings as those provided by otter haven projects, they will also play with their human benefactors. They seem to love sliding down river banks in the mud or snow and also enjoy playing with pebbles. Zoologists still don't understand why some animals are more playful than others, but play is demonstrably an important part of the otters' life; and their apparent sense of fun certainly makes them one of our favourite mammals.

The type of food the cubs learn to catch consists mostly of fishes – trout, sticklebacks, and so on – but they also hunt water birds and water voles and various other creatures such as crayfishes and frogs. On the coast they often catch crabs.

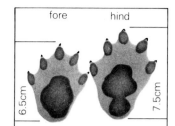

fore   hind

6.5cm   7.5cm

EUROPEAN OTTER (*Lutra lutra*) Also called water weasel, water dog, burn dog, tek
Size male weighs 10.3kg (23lb) and total length is 1.2m (4ft); female weighs 7.4kg (16lb) and is 1m (3ft 3in) long
Colour rich brown, often paler throat and lip markings
Breeding season all year
Gestation period 62 days
No of young 1-5; usually 2-3
Lifespan unknown in wild otter; 20 years in captive North American otter
Food fish, amphibians and crustaceans. Also small mammals and birds
Predators No natural predators
Distribution in England mainly restricted to south west, East Anglia; also parts of Wales; more numerous in Scotland

No haven for the otter

It is clear that the European otter is vanishing from many of its haunts – chiefly because of man-made changes to its environment. Polluted rivers, for example, will affect the food supply and detergents in the water can destroy the otter's indispensable waterproofing; once its diving suit is damaged, the otter can no longer resist the wet and cold and so succumbs to both and dies. Its numbers have declined so much in England and Wales that it is now a protected species. Positive steps have been taken to protect its environment and promote its increase by the setting up of nationwide otter haven projects. There is still a great deal of argument over whether the mink – which escaped from fur farms and now lives in many areas in the wild – competes to a serious extent with the otter. Scotland is undoubtedly the best place to see otters today.

Typical otter country in Wester Ross, Scotland – an undisturbed riverbank and plenty of natural hollows behind the boulders, providing ideal breeding sites for the secretive otter.

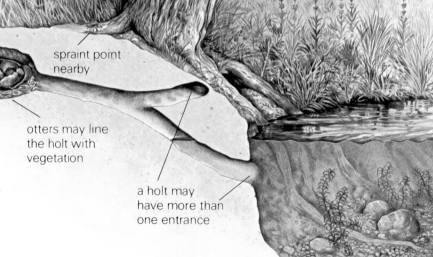

the chamber is well above the winter flood-level

spraint point nearby

otters may line the holt with vegetation

a holt may have more than one entrance

The breeding holt

Otter holts are always well-hidden. The female needs to avoid drawing attention to the holt, especially when the cubs are very young. Thick bankside vegetation or an underwater entrance allows the otter to enter or leave the holt unseen.

AGGRESSIVE WATER SHREWS

If you hear intermittent, high-pitched squeaking and twittering from the grass as you walk along a river bank, the chances are that the noises are not made by a cricket or bird but by a water shrew that is probably engaged in some aggressive act.

Below: Water shrews have many predators including foxes, stoats, weasels, kestrels and owls. Pike and perch eat them too. Domestic cats also kill shrews but then do not always eat them because of their strong musky smell.

Water shrews are one of the five species of shrew found in the British Isles, and they differ from the others in several ways. They are the largest, at almost twice the weight of the common shrew and four times the weight of the pygmy shrew. Even so, they weigh little more than a 50 pence piece. In addition, water shrews have rather long fringes of greyish hair on their tails and on the edges of their hind feet, whereas other shrews have shorter fringes on their feet. Their coats are glossy black above and greyish, sometimes white, below.

Watery habitat The water shrew is widely distributed and has been recorded over most of mainland Britain, on the Inner Hebrides, Hoy in the Orkneys and the Isle of Wight. As their name suggests, water shrews are at home in water. They usually live close to a stream, pond or slow flowing river, preferably where there is good cover. They are secretive and shy, and hide from larger mammals and birds.

Wherever they are, water shrews establish a territory where they make a system of runways and tunnels. Dr Peter Crowcroft, who did much of the pioneering work on shrew biology, categorised these areas into three types: surface runways which enable the shrew to travel about swiftly; surface burrows which are used for finding food, rather like a mole's burrow, and extend through the grass in a great maze, connecting with the underground tunnel systems; and tunnels, used for

The diving water shrew

WATER SHREW
(*Neomys fodiens*)
Size Body up to 9cm (3½in),
tail up to 7cm (2¾in).
Colour Dark, almost black,
above, may have brownish
tinge; usually pale beneath.
Breeding season Usually
April-October.
Gestation About 24 days.
No of young 3-8, with 2-3
litters a year.
Lifespan 2-18 months.
Food Invertebrates
underwater as well as on
land; small fish, frogs which
are paralysed or killed by a
venomous saliva.

nesting and for access to the water. Water shrews can dig very effectively, using their snouts as well as their feet. They will not tolerate other shrews in their territory – pandemonium breaks out if a stranger ventures in by mistake, but they share the burrows of voles and moles.

Shrews remember the geography of their burrows and runways in great detail, unlike many other mammals which rely on scent markings to direct them, so that by the time they have run through a burrow for a second time they travel on 'auto-pilot'. If an obstacle is removed from their path, like a clump of grass or a root, they negotiate as though it were still there. Conversely, if an object which

Above: The coat of the water shrew is very water repellent, and the shrew floats high like a cork. As it dives, the fur traps a film of air bubbles, and the relatively large amount of trapped air compared with its weight makes it awkwardly buoyant. With fast, frantic strokes of its feet it scrabbles to the bottom in search of food. It has to paddle all the time as it noses around the stones and weeds because the moment it stops it will shoot up to the surface again (below).

other mammals might avoid, such as a prickly sweet chestnut, falls into a runway, shrews will run headlong into it.

Agile movers The fringes of hair on the water shrew's tail and hind feet help it to propel itself through the water. The feet give extremely good grip which is useful for overturning stones at the bottom and in tree climbing. Water shrews are versatile animals and are not restricted to aquatic and ground habitats. When they locate an insect in a tree or bush, they climb after it. If required they can also use their tails to grip twigs or branches.

All small warm blooded animals need to work hard to maintain their body temperatures in a climate like that in the British Isles. Water shrews forage intently to eat enough to sustain the high metabolic rates which dominate their short, active lives.

Their body temperature is about the same as our own, but they have to work about twenty times as hard to maintain it. A shrew may eat just over its own body weight in 24 hours. As they do not hibernate, it is hardly surprising that they seldom live more than 18 months at a maximum. Constant high activity places enormous demands on the heart and circulation.

Touching food No one has described the way shrews find their food better than Crowcroft: 'When first one watches shrews feeding, one is struck by their inability to find food until they almost fall over it.' This may seem surprising in view of the shrew's apparently well developed snout. Smell and sight are important, even though shrews are extremely shortsighted; but touch is paramount in the search for food, although it can lead to disaster. The shrew's fast reactions help it to escape easily from some of its more dangerous encounters. When a water shrew meets a mole in its tunnel system, the mole barely has time to flail out with its legs before the shrew is on its way.

Aggressive habits The word 'shrewd' derives from 'shrew' and originally meant villainous before its present day meaning evolved. Water shrews are possibly the most aggressive of all the shrew species and will not tolerate the presence of other shrews, regardless of their age, sex, or species. Serious fighting, however, is minimised by a strict set of rules.

If one shrew meets another in a runway the intruding shrew usually backs off fast and scurries away. If the stranger will not withdraw the two engage in a squeaking and screaming contest where the one who screams longest and loudest causes the other to give way. The fight escalates if this does not resolve the dispute.

The first stage of water shrew aggression involves swift biting of the opponent's tail. The second stage of this trial of strength is continued screaming with one or both animals rearing up on their back legs. This sudden increase in apparent size, and the flashing of white belly fur on some animals usually settles the argument, but flashing may continue for several seconds.

During the third stage of agonistic behaviour the shrews grip and grapple with each other with an increasing lack of mercy. The last resort is for the shrew to fling itself on its back, striking out with its legs and screaming at the top of its voice. The effect on the opponent is devastating, but can be lost if both act together because neither can see what the other is doing.

Research has shown that the order of these stages of aggression is extremely strict. Each stage is more energetic than the last, so by trying one stage at a time the chances are that the fight will break up sooner rather than later, unless the two are very evenly matched. That way not too much energy is wasted.

Blind alley nests Nests are made within the burrow system, also according to a strict procedure. The shrew collects grass and other soft fibrous materials which it dumps outside the nest chamber, a blind alley off one of the burrows. It squeezes into the chamber, drags the material inside, then turns round in circles, gradually releasing its grip on the grasses. The resulting hollow ball is then strengthened with further building work.

Shrews use their nests to rest for short

Above: A watercress bed on the Itchen at Alresford in Hampshire, a typical habitat of water shrews.

Below: Water shrews eat invertebrates like earthworms, as well as frogs, small fishes and plant material.

periods and to raise their young. The peak breeding season is late May and June. Pregnancy lasts about three weeks, and the young are born blind and hairless. A water shrew's nest should never be disturbed: the young will quickly chill and the mother may bite the intruder. Each baby weighs about half as much as a penny coin at birth. After five weeks they reach full size and are unceremoniously ejected from their mother's territory. An early litter may breed the same year, but the majority do not. Many first year shrews perish in their first winter: the twin blights of cold and starvation are against them. Predators also take their toll.

Young shrews moult in October, growing a thicker coat before winter sets in. Shrews do not hibernate since their metabolic makeup does not allow it. Their parents will be lucky to live until December, as they do not grow a winter coat the second time round. Their teeth have worn down to stumps and the cold catches up with them. Maintaining their almost incessant activities, which include diving into near-freezing streams, becomes too much, and they usually die of starvation. Meanwhile the next generation will be scurrying around their runways and burrows; the only clue they are there is their high pitched twittering and squeaking.

SUMMER AND WINTER PIPISTRELLES

Contrary to popular belief, pipistrelle bats often emerge from hibernation on mild winter nights when insects are easier to catch.

Above: The River Thames at Marlow, a typical pipistrelle habitat. Scientists discovered that pipistrelles only emerged in winter when the temperature was 8°C (46°F) or above. On colder nights they could not catch enough insects to make feeding profitable. From data collected over 350 winter nights they found that on 76% of nights the bats did what was expected, showing that bats are able to choose the most advantageous nights to emerge.

Pipistrelle bats, the commonest British bats, occur throughout the British Isles. Like most of the 15 British bat species, pipistrelles have two types of roosts: a nursery roost which females occupy during summer, and a winter roost or hibernaculum, occupied by bats of both sexes during the colder months when insect food is scarce and when bats may spend long periods with their body temperature close to that of their surroundings.

Summer feeding Female bats arrive at their nursery roosts in spring, and although adult males are not allowed into these maternity wards, leading a solitary existence in summer, young males may be tolerated before they reach puberty. Females give birth in mid-summer, generally to a single young, but occasionally to twins.

Each night, shortly after sunset, the females leave their roost to forage. They are hungry, since they have not eaten for almost 20 hours. A large colony takes up to an hour to leave the roost and the animals can be heard squeaking as they jostle and push to the exit hole. A single bat often leaves first, a few minutes before the others, who follow in bursts, with up to 40 bats emerging in a 30-second interval, followed by a lull when no bats emerge. This emergence in outbursts may confuse predators such as a waiting cat or owl which will find it difficult to select an individual for attempted capture.

To discover how far the bats travelled on their foraging flights, a group of scientists caught them in nets as they emerged from the roost, and attached gelatin pill capsules to the fur on their backs with surgical cement. The capsules were filled with a chemiluminescent mixture which glows bright green for a few hours so that the bats look rather like fireflies. The capsule is groomed off the fur by the bats when they return to their roost.

With many helpers looking for these 'fireflies', it is possible to get a good idea where the bats are foraging. Researchers can then return to this area night after night with powerful torches which enable them to catch a glimpse of the reflective tape stuck to the tiny numbered rings attached to the bats' forearms (under the appropriate licences). They found that pipistrelles prefer to forage around riverside vegetation, whether just above the water or around the deciduous trees growing along the riverbank. In a rich lowland valley in northeast Scotland, the bats travelled up and down the river system, straying from it only to feed around nearby ponds and foraging up to

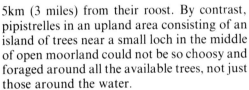

Above and below: Pipistrelle bats. A male pipistrelle will weigh 6.25g (just under ¼oz) at the beginning of October, its fattest time. It then loses weight steadily through the winter until the end of April when it weighs only 3.75g (⅛oz). In other words, it loses about a third of its weight during this time by gradually using up its fat reserves as winter progresses. By the end of winter, with fat reserves at their lowest, mortality is at its highest.

5km (3 miles) from their roost. By contrast, pipistrelles in an upland area consisting of an island of trees near a small loch in the middle of open moorland could not be so choosy and foraged around all the available trees, not just those around the water.

Using different colour reflective tape for different individuals scientists discovered that pipistrelles often feed at the same time and place each night. By sitting outside a bat roost throughout the night and counting the bats coming and going it is possible to work out the activity patterns of the bats. During pregnancy their nightly feeding lasts from one to five hours and females return to the roost throughout the night after feeding. After birth, the demands of a hungry infant change the pattern of foraging because the mothers return to the roost in the middle of the night to suckle their young. Their second flight coincides with the dawn peak of flying insects.

Independence As soon as the young bats can fly (towards the end of August) the mothers leave the nursery to roost, and their young remain. This unusually early severance (for mammals) of the bond between mothers and young may reduce competition for available food. In addition females look for mates during September since all bats living in temperate latitudes mate in autumn and spermatozoa are stored in the uterus until spring when ovulation occurs and pregnancy follows. Attempts to understand how bats achieve this remarkable feat have prompted much research; in particular, researchers want to find out how active bats were during the time they stored spermatozoa.

Food in winter During summer it is probably quite easy for bats to find enough to eat, but in winter there are few live insects. So what must bats do to survive? They could avoid the

can allow their body temperatures to drop right down to about the same temperature as their surroundings. So instead of around 37°C (98.4°F) a bat in a cold place could have a body temperature of only 5°C (41°F). During this period the bat is said to be in torpor, and bats enter torpor for most of the winter. This prolonged period of torpor is called hibernation.

A hibernating bat is vulnerable because at such a low temperature it cannot move, so it must choose a safe place, free from predators, in which to spend the winter. On the other hand, the advantage of hibernation is that the bat uses less energy while it is in torpor than it would if it kept up a high body temperature. This is how bats manage to live on their fat reserves: hibernation allows them to use their stored fat at a slow rate. No-one knows exactly how bats and other hibernators can survive at low temperatures. Certainly it is an ability that man does not have. Hill-walkers who become lost and die of exposure, or old people who die of hypothermia, do so because their body temperature has dropped and they cannot warm themselves up on their own. But this is just what the bat can do; apparently bats can become torpid and then become active again by warming themselves up at any time.

Winter arousal The warming-up process, called arousal, uses up much energy so hibernators arouse in winter comparatively

cold months by migrating to warmer countries–this is what almost all insect-eating birds do (for example, swallows and swifts). In other parts of Europe it has been shown that bats, too, migrate south in autumn and again in spring, but at present experts think that British bats stay in this country throughout the winter.

Since insect food is rarely available in winter, bats live on their fat reserves, and pipistrelles become very fat during autumn. They lose weight steadily through the winter as they use up their fat. However, these reserves, important though they are, do not alone keep the bats alive at this time. Like hedgehogs and hamsters, for instance, bats

Above: Pipistrelles are crevice seeking bats and make their nursery and winter roosts in narrow spaces. In summer they roost between slates and roofing felt, or between wooden rafters and brickwork. Many bats roost together in these situations–occasionally as many as a thousand in one place. In winter their favourite hibernacula are in medieval country churches in gaps in mortice joints between the beams.

Pipistrelle summer haunts at dusk

It is dusk, and the bats have emerged to forage for insect food. They fly up and down on a beat for about 20 minutes, catching swarming insects, then they fly directly to another site and establish another beat. Pipistrelles prefer to forage around riverside vegetation just above the water or around the deciduous trees along the river. They take small non-biting midges, caddis flies and lacewings. On nights when there is a hatch of mayflies these are selected. The bats are seldom aggressive to each other, unless the number of flying insects falls when they defend their chosen feeding sites against intruders. If insects are plentiful it is common to see 20 to 30 pipistrelles flying up and down, and never colliding with each other.

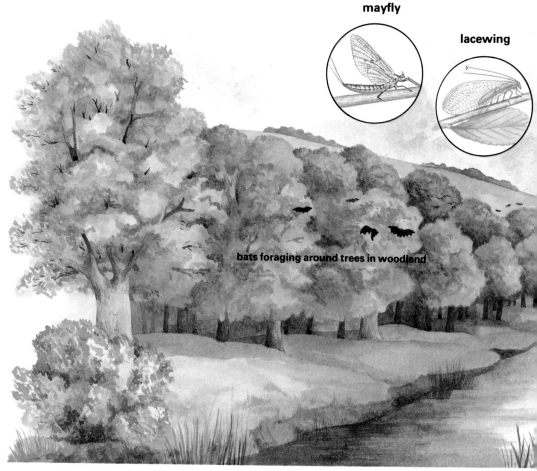

mayfly

lacewing

bats foraging around trees in woodland

rarely. An undisturbed bat may only arouse once a fortnight through the winter. However, arousal can be triggered if the bat is disturbed, so it is important that the hibernation sites of bats are protected.

Although hibernation and fat storage are the main bat survival strategies for winter, many people have seen bats flying during this time. Scientists studied pipistrelles in East Anglia for three winters to find out how often they left their roosts. They were flying outside their roosts on almost half the nights between October and April. Why were they active? And what were they doing?

Using bat detectors researchers were able to show that the pipistrelles were feeding during the winter nights. The bat detector picked up the very high frequency sounds with which the bats detect insects and it made them audible. Thus researchers could find the bats in the dark by walking around until they heard the characteristic 'pitter-patter' noise from the bat detector. They could also measure the rate at which bats feed. As a bat dives towards an insect it is about to catch, the 'clicks' of high frequency sound it emits become very close together and sound like a continuous buzz on the bat detector. So by following individual bats the researchers could listen to them through the bat detector to find out if they were feeding or not—and found that they were. The bats' feeding rate could be measured by recording how

Above: The pipistrelle bat always catches its food in flight. It takes small items directly into its mouth, but larger items may be caught in its very flexible wings. The bat then flicks the food from its wing into the membrane between the back legs from where it takes the food with its mouth. Dawn and dusk are the main feeding times, although a pregnant bat may feed for up to five hours before returning to the roost. Nursing mothers return to suckle the young.

often feeding buzzes were heard: the warmer the temperature, the more buzzes occurred. So the scientists could predict when the bats ought to come out to feed. They collected data for over 350 nights and found that bats were active on warmer nights and stayed in their roosts on colder nights. The work has shown that, in addition to storing fat and becoming torpid, pipistrelle bats feed in winter. It is worthwhile for them to use energy to forage then because they are able to choose the warmer nights when they are likely to catch most insects.

Steep declines in bat populations are reported in England, but there is no evidence of a major decline in Scottish populations.

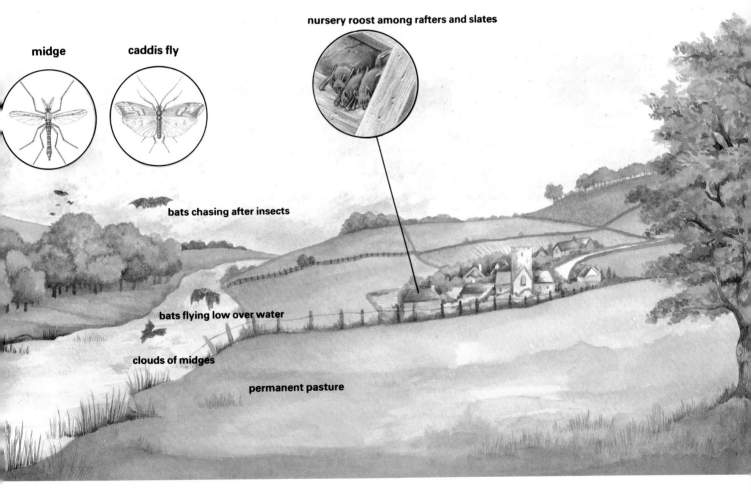

midge

caddis fly

nursery roost among rafters and slates

bats chasing after insects

bats flying low over water

clouds of midges

permanent pasture

MINK: EXPERT ESCAPERS

American mink were introduced into Britain early in this century to be bred in captivity on fur farms but they soon showed a talent for escape.

The American mink, which is now widespread in mainland Britain, is often mistaken for a small otter, particularly when it is wet and its normally bushy tail appears thin. However, the mink is much smaller than the otter, less than one tenth of the weight and only half the length. Both animals have rich, glossy fur and, when wet, appear dark in colour. However, the fur of the mink generally dries to a darker brown than that of the otter, although very pale individuals occur from time to time.

The mink has a blunt snout, and its ears are small and almost hidden in fur. The neck is thick and muscular, the body slender and sinuous. The legs are short, and the tail well

Below: The mink has several dens in its territory, where it rests during its various travels. These can be sited in a variety of places – in a hollow tree trunk, among boulders or even just in a heap of dry grass or bracken. As it moves from end to end of its territory, the mink leaves strong-smelling droppings which serve to mark the territory and advertise the presence of the owner.

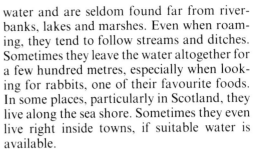

AMERICAN MINK
(*Mustela vison*)
Size Male about 1kg (2¼lb), 60cm (24in). Female about 600g (1¼lb), 50cm (20in).
Colour Usually dark brown, occasionally pale. White spots or flecks on chin, throat and on underside.
Breeding season Young born April to May.
Gestation period 39 days, but variable delayed implantation can extend this to up to 76 days.
Number of young Average 5-6. One litter a year.
Lifespan Short in wild, but occasionally animals live for several years.
Food Fish, small mammals and birds (especially eels, rabbits and waterfowl). Occasionally crayfish.
Predators Man. Occasionally killed by otters.
Distribution See map on page 189.

Below: Mink seem to have much less fear of people than most wild animals, and they are frequently seen at quite close range. They have even been known to try and steal fish and sandwiches from fishermen on river banks.

furred and bushy when dry.

From the New World to the Old The mink which now live in Britain came originally from North America, where they are found in the temperate forest regions of the United States and Canada. When they were imported early in this century, to be bred on farms for their extremely valuable fur, they were not tame in any way, remaining as wild as their native forebears. They soon showed themselves to be expert escapers, and by about the middle of this century had escaped in sufficient numbers to start breeding in the wild.

Feral mink, also descended from American stock, are found in many other European countries such as Iceland, Norway, Sweden, Denmark, Finland, Spain, North Germany and Russia.

European relatives Not to be confused with the American species is the native European mink. It looks very like its American cousin, but is generally smaller and lighter, and has a distinguishing white spot on its upper lip. It is not found in Britain, and is present mainly in Eastern Europe, Finland and Russia. It was probably once more widespread in Europe, and small populations still persist in northern and western France. The European mink is not as aggressive and adaptable as the American species and, far from expanding its range, seems to be on the decline.

Waterside habitats Mink like to live near water and are seldom found far from riverbanks, lakes and marshes. Even when roaming, they tend to follow streams and ditches. Sometimes they leave the water altogether for a few hundred metres, especially when looking for rabbits, one of their favourite foods. In some places, particularly in Scotland, they live along the sea shore. Sometimes they even live right inside towns, if suitable water is available.

If you see something like a large weasel or small otter, near a lake or river, or on the sea shore, it may well be a mink. Unlike the otter, which is active only at night when there is no danger of human disturbance, the mink is about at all hours, even when people are in evidence.

It is difficult to estimate the number of mink in Britain today. A mink needs several miles of waterside to make its home and, considering the thousands of miles of watercourse throughout Britain, there must be thousands of mink in the country.

Long, narrow territories Mink are territorial animals. A male mink will not tolerate another male within its territory, but appears to be less aggressive towards females. Generally, the territories of both male and female animals are separate, but a female's territory may sometimes overlap with that of a male. Very occasionally it may be totally within a male's.

The territories, which tend to be long and narrow, stretch along river banks, or round the edges of lakes or marshes. Sizes vary, but they can be several miles long. Female territories are smaller than those of males.

Each territory has one or two central areas (core areas) where the mink spends most of its time. The core area is usually associated with a good food supply, such as a pool rich in fish, or a rabbit warren.

The mink may stay in its core area, which can be quite small, for several days at a time, but it also makes excursions to the ends of its territory. These excursions seem to be associated with the defence of the territory against possible intruders. It is likely that the mink checks for the presence of any strange mink, and leaves droppings (scats) redolent of its personal scent to reinforce its territorial rights.

In the territory, there are several dens. These may be in the roots or trunk of a waterside tree, or among boulders, or they may be just temporary beds of dry grass or bracken. The mink may even use the abandoned nests of large birds, for it is an agile climber.

Some mink seem to live temporarily without territories. These animals (called transients to distinguish them from territorial residents) are males which have left their territory to seek out females; young mink which have left their mothers and are looking for territories of their own to settle in; or animals which are seeking a better territory than the one they have abandoned. (If a

territory proves for any reason to be unsatisfactory – perhaps not rich enough in food – the owner will abandon it and set off to find something better.)

Mink are solitary animals for most of the year. Male mink avoid contact with other males at all times, and seek out females only in spring. Females have more contact with their own kind for, apart from meetings with males for mating, they have the company of their offspring for a few months until they are old enough to go their own way.

A varied diet The mink eats anything big enough to be worth its attention and small enough to be caught and overcome. That means anything from about the size of a mouse up to the size of a rabbit.

Although mink can swim well and catch many kinds of fish (showing a marked preference for eels), they are not so well adapted for aquatic hunting as the otter, and are not so dependent on fish for food. They eat more fish in winter when the fish swim slower and are easier to catch.

They are equally fond of furry prey and take mice, voles, rats, squirrels and young rabbits. If these are scarce, they turn to birds. As might be expected from an animal with such aquatic inclinations, the mink takes mainly waterfowl, particularly coots and moorhens.

A serious pest? Much is made of the mink's occasional raids on domestic stock and game,

but studies have shown that the extent of these depredations is usually slight and of little importance countrywide, although infuriating to the farmer or gamekeeper who has suffered. Similarly, accusations have been made that mink have in some places exterminated such waterside creatures as moorhens and water voles. However, other places are known where flourishing populations of mink and voles or moorhens exist side by side, so the matter still rests in doubt.

The catholic tastes of the mink mean that, by and large, it is not in any great competition for food with the otter, and is unlikely to have contributed to its decline. Fish stocks, the staple food of otters, have suffered no apparent set-back following the appearance of the mink.

Nevertheless, the mink's destructive habits have made it influential enemies, and it has been officially branded as a pest. Local control may sometimes be necessary, and in such cases the pest control officers of the Ministry of Agriculture, Fisheries and Food can be extremely helpful.

Social calls Male mink get the mating urge from February onwards. They do not have permanent mates, so they leave their territories and travel the countryside in search of females.

Mink show the curious phenomenon of delayed implantation. Although the true gestation period is 39 days, the embryo may stop developing for a variable period, so that as long as 76 days may elapse before the litter arrives. Between 45 and 52 days is average.

Usually a female has 5 or 6 cubs. The male does not help rear the young, and their feeding and training are entirely the responsibility of the female. The young remain with their mother until the autumn, when they are fully grown. They then leave her to find territories of their own. It seems likely that there is a considerable rate of failure in this move to independence, and first year mortality, due to starvation, is high.

Unsuitable pets Although mink are not too difficult to catch with a special trap, they are not to be recommended as pets since they are fierce and almost impossible to tame.

Above: When hunting for fish, mink prefer to move along the bank, or on boulders, fallen logs or low-hanging branches, peering into the water. When they see a fish, they dive in to pursue it.

Opposite page: A mink emerges cautiously from its hideout in an old quarry to go on a hunting expedition.

forefoot

hind foot

dropping

Above: Mink footprints and dropping. Toe marks do not always show up in the wild.

Above: Mink are widespread in mainland Britain, except in the mountainous regions of Scotland, Wales and the Lake District. They are also found in the isles of Arran and Lewis. In Ireland they are less common.

The entries listed in **bold** type refer to main subjects. The page numbers in *italics* indicate illustrations. Medium type entries refer to text.

ACKNOWLEDGEMENTS

Photographers' credits Heather Angel front cover, 6-7, 10 (bottom), 11 (top), 12, 15, 16, 17, 26 (top), 28, 29, 32 (bottom), 35 (middle), 36 (bottom), 38, 47 (top), 48 (bottom), 59, 65, 68, 70 (top), 72-3, 77 (top), 78 (top), 85, 89 (bottom right), 92, 95, 100 (top), 102, 103, 106, 107, 111, 113, 116 (bottom), 117, 119, 120-21, 122 (top), 125, 129, 130 (bottom), 131 (bottom), 132 (bottom), 134 (middle), 136, 139 (bottom), 143 (bottom), 150, 165 (top), 188: Aquila Photographics/ JB Blossom 154 (bottom), 156 (top); R Kennedy 152 (bottom); W Lankinen 154 (top), 156 (bottom); M Leach 168 (top); T Leach 86; DI McEwan 51 (bottom), 134 (top); RG Powley 162; N Rodney Foster 160 (bottom); EK Thompson 168 (bottom); MC Wilkes 166: A-Z Collection 56, 58; Ian Beames 52 (bottom), 145, 148: Biofotos/ BM Rogers 161: Bob Gibbons Photography/ R Fletcher 44-5, 90; Bob Gibbons 13, 18, 50 (bottom), 71, 75, 96-7; Peter Wilson 36 (top): Ursula Bowen 22: Bruce Coleman Ltd/J Burton 100, 108, 116 (top), 174-5; Jane Burton 124, 180, 187; Bruce Coleman 182, 185; B Langsbury 151; L Lee Rue III 186; J Markham 127; Prato 87 (bottom right); Hans Reinhard 98, 112, 114, 118: Michael Chinery 20 (bottom), 35. JPA Clare 171 (bottom): John Clegg 8, 9, 10 (top), 11 (bottom): Eric Crichton 62, 64, 67: Keith Easton 115: Ross Gardiner 37: Dennis Green 24 (bottom), 25, 80, 153 (middle), 155: Brian Hawkes 149: Nigel Holmes 40 (top), 42 (top), 43 (top), 45 (bottom), 47 (bottom), 49, 50 (top), 53 (top, bottom right),

54-5, 79, 88: RW Ingle 122 (bottom), 123 (bottom): G Kinns 176, 177 (bottom), 178: London Scientific Films 181 (top): TT Macan 131 (top): John Mason 89 (bottom left), 91, 137 (top), 143 (top), 184: S & O Mathews 30, 31 (top): RT Mills 24 (middle), 31 (top), 157, 163, 167 (middle): M King & M Read/M King 43 (middle), 46 (top), 76, 77 (bottom), 82 (bottom); M Read 40 (bottom), 42 (bottom left), 81: Pat Morris 26 (bottom), 32 (top), 159 (bottom): Natural History Photographic Agency/J & M Bain 78 (bottom); A Barnes 27 (top, bottom right); NA Callow 126 (bottom); L Campbell 74, 87 (top); DN Dalton 160 (top), 169; S Dalton 133 (middle), 135; J Goodman 99, 110; B Hawkes 165 (bottom); EA Janes 61 (bottom); M Leach 146-7; B Newman 128; P Scott 152 (bottom): Nature Photographers Ltd/SC Bisserot 138, 167 (bottom), 183, 185; FV Blackburn 69, 71 (top), 164; D Bonsall 101; Idris Bowen 23; N Brown 43 (bottom); B Burbidge 24 (top); NA Callow 140 (top); K Carlson 158, 181 (bottom); A Cleave 31 (bottom), 82 (top); JV & GR Harrison 20 (top), 70 (bottom); MR Hill 48 (middle), 153 (bottom); D Hutton 172, 173 (top); EA Janes 46 (bottom); C & J Knights 35 (top), 130 (top); A Mitchell 66; D Smith 34; P Sterry 27 (left), 133 (bottom), 134 (bottom), 139 (top), 142; R Tidman 31 (middle): Naturfoto/G Harrison 137 (bottom); L Jonge 171 (top): John F Preedy 179: Premaphotos Wildlife/ KG Preston-Mafham 21, 42 (bottom right), 48 (top), 52 (top), 53 (bottom left), 57, 132 (top), 140 (bottom): Press-tige Pictures/D Avon & T Tilford 19, 123 (middle), 183 (top): Richard Revels 83, 126

(top): John Robinson 14, 93, 105, 109, 141, 144, 173 (bottom): BS Turner 159 (top): Wildlife Services/M Leach 51 (middle), 177 (top), Gerald Wilkinson 61 (top).

Artists' credits Stephen Adams 120: Graham Allen/Linden Artists 180: Fred Anderson 113: Norman Arlott 146, 149, 158, 159, 170: Artists' Partners/Claire Askaroff 184-5; Bob Bampton/The Garden Studio 33 (insets): Russell Barnett 22-3 (insets): Linsay Blow 39 (line): Robert Burns/ Drawing Attention 17: Martin Camm/Linden Fleury 51 (line): Wayne Ford 29 (insets), 153, 155, 161, 163, 164, 167, 168, 173: Philip Gibbs 38: Hayward Art Group 9, 60 (leaves), 63 (leaves, catkins), 72, 84, 85, 91, 94-5, 109, 117, 137, 142, 143, 178, 189 (line): Richard Lewington/The Garden Studio 139, 140, 168 (line): Kathryn Lunn 27: David More/Linden Artists 54, 57, 58, 60 (trees), 65, 67, 70: Orbis Publishing Ltd 151: Denys Ovenden title page, 98-9, 103, 104, 105, 106, 115, 119, 123, 125: Sandra Pond 80, 83: Adrian Rigby 155 (line): Andrew Riley/The Garden Studio 130: Gordon Riley 133, 135: Colin Salmon (maps) 15, 22-3, 29, 32 (diagram), 33, 80, 87, 115, 153, 161, 172: Joan Sellwood/The Garden Studio 174, 189: Ed Stewart 39 (map): Carole Vincer 91 (top far left), 92: Phil Weare/Linden Artists 75.

Index compiled by Richard Raper of Indexing Specialists, Hove, East Sussex.

Typesetting PHOTOCOMP LTD, BIRMINGHAM; Printing & Binding PRINTER INDUSTRIA, GRÁFICA S.A. BARCELONA;
Separations YORK HOUSE GRAPHICS, HANWELL; COLOURSCAN OVERSEAS CO PTE LTD, SINGAPORE;
Paper PERIGORD-CONDAT, FRANCE